NE ASSOCIATION, INC.

# OVERPOPULATION OF CATS AND DOGS

# Proceedings of a Conference

New York City
September 11 & 12, 1987

Sponsored by

# Overpopulation of Cats and Dogs

## Causes, Effects, and Prevention

Edited by

Marjorie Anchel, Ph.D.

Fordham University Press

New York

1990

© 1990 BY NEW YORK STATE HUMANE ASSOCIATION, INC.
ALL RIGHTS RESERVED

FOR INFORMATION AND PERMISSION TO QUOTE OR REPRINT,
WRITE THE COPYRIGHT-HOLDER:
NEW YORK STATE HUMANE ASSOCIATION, INC
P.O. BOX 284
NEW PALTZ, NEW YORK 12561

LIBRARY OF CONGRESS CATALOG CARD NUMBER
90-82350

ISBN 0-8232-1296-3

FIRST EDITION
90    1 3 5 7 9 10 8 6 4 2

PRINTED IN THE UNITED STATES OF AMERICA

GRAPHICS AND PAGE LAYOUT DESIGN
BY
THE INFO DEVELS INC

# CONTENTS

| | |
|---|---|
| FOREWORD | vii |
| EDITOR'S PREFACE | ix |
| ACKNOWLEDGMENTS | xi |
| STATE OF NEW YORK LEGISLATIVE RESOLUTION | xiv |
| THE SPEAKERS | xvii |

## DAY ONE

OPENING OF CONFERENCE AND WELCOME .......................... 3

- Marjorie Anchel, Ph.D.
- Morton Hillman, Member, NYS Assembly
- John Kullberg, Ed.D.
- Martin Kurtz

KEYNOTE ADDRESS ................................................................. 7

- Christine Stevens

ANIMAL CONTROL: HOW IS IT BEST HANDLED?
RURAL, URBAN, AND METROPOLITAN APPROACHES .................. 15

- Margaret Geraghty
- Robin Remick
- Gretchen Wyler

WHO IS RESPONSIBLE FOR PET OVERPOPULATION? ................... 41

- Barbara Cassidy

CONTROL OF FERAL CATS ....................................................... 57

- AnnaBell Washburn

HELL IN PARADISE: VIEQUES, PUERTO RICO ............................... 73

- Ann Cottrell Free

EUTHANASIA: AGENTS USED; PUBLIC PERCEPTION;
STRESS ON TECHNICIANS ........................................................ 77

- Gordon Robinson, V.M.D.
- Ingrid Newkirk

## Day Two

RESPONSIBILITIES OF SHELTERS ............................................. 97
- Barbara Cassidy
- John Kullberg, Ed.D.
- Kathleen Young

WHAT'S WRONG WITH POUND SEIZURE? ................................ 119
- John McArdle, Ph.D.

WHAT PART CAN VETERINARIANS PLAY? ............................... 133
- Arthur Baeder, D.V.M.
- Robert Case, D.V.M.

NEUTERING SEXUALLY IMMATURE ANIMALS ........................... 147
- Leo Lieberman, D.V.M.

CHEMOSTERILANTS ............................................................ 163
- Wolfgang Jöchle, Dr. Vet. Med.

EDUCATION OF CHILDREN AND OF THE PUBLIC ....................... 171
- Ronald Scott
- Sheri Trainer

ATTITUDES TOWARD NEUTERING AND EUTHANASIA .................. 189

MEDICAL MYTHS
- David Samuels, V.M.D.

PSYCHOLOGICAL AND RELIGIOUS INFLUENCES
- Murry Cohen

VIEWPOINTS OF ANIMAL RIGHTS ADVOCATES
- Ingrid Newkirk
- Tom Regan, Ph.D.

LEGISLATION: PAST AND FUTURE ........................................... 217
- Elinor Molbegott, Esq.

CLOSING OF CONFERENCE .................................................. 231
- Samantha Mullen, NYSHA Administrator
- Maurice D. Hinchey, Member, NYS Assembly

APPENDIX A—UPDATES ...................................................... 235

APPENDIX B—EDITOR'S NOTES ............................................ 243

# Foreword

The New York State Humane Association conference on "Overpopulation of Cats and Dogs" of September 1987 brought together a diverse group of speakers who were willing to share their thoughts and ideas.

It is important that when we work *for* animals, we work *with* people. Even though we have our own agenda and see things from our viewpoint, it is when we listen to the other person that we can all do a better job and meet our goals of reducing the numbers of surplus animals.

In the following pages you will read materials with which you will both agree and disagree. Perhaps you will think some ideas are impossible, but you will learn here of ideas that are working. It is up to you to make the choice of trying something new or different and giving it an opportunity to improve your programs, methods, local and state legislation, and your impact in the community.

When you finish reading these presentations, you will know that you are hearing from people who are in the trenches: those who have to and do deal with the day in, day out problems of surplus animals or have studied the problems and have found solutions that may help to resolve them.

The statements you read can and should be used to make *more* people aware of local problems and solutions. Although you may have said the same thing one hundred times, quoting any of the speakers may get you a new audience.

It is now up to each of us who read these proceedings to share with the public the dilemmas that we face each day as we try to cope with the surplus animal problem, and put forth new ideas to change public attitudes. The following pages will give you that unique experience.

After you have read this material, buy another copy for your library, your veterinarian, your local animal shelter. Each time we share our knowledge we cause a ripple effect: every person who becomes involved helps share the burden. My congratulations to all of the outstanding speakers who took time to share their knowledge, and to the New York State Humane Association for making a difference through this conference.

Phyllis Wright
Vice President/Companion Animals
The Humane Society of the United States

# Editor's Preface

The speakers at the NYSHA conference represent a diversity of backgrounds, training, professions, and experience. They met to discuss one subject: overpopulation of cats and dogs. Each had a special viewpoint; each contributed a special expertise. Among them are animal control and shelter management experts, legislators, veterinarians, humane educators, a psychiatrist, an attorney specializing in animal law, a scientist specializing in birth control methods for animals, animal welfarists, and leaders of the animal rights movement.

Each speaker, besides having a unique point of view, has a special style of presentation, reflected in the use of language, the relationship to the audience, and the formality or informality of approach. In editing the talks, I have tried to preserve the "flavor," so that the reader will feel that, to some extent, he is "hearing" the speakers.

Many who attended this conference came away, not only with new ideas, but in some instances with a changed viewpoint. I hope that these printed proceedings will reach a larger audience, and will result in a more widespread understanding of the problem of overpopulation of dogs and cats, a more general acceptance of different approaches, and in increased cooperation of veterinarians, legislators and humane groups in seeking and implementing more effective and more humane solutions.

Marjorie Anchel, President
New York State Humane Association, Inc.

# ACKNOWLEDGMENTS

The New York State Humane Association expresses its deep gratitude to those who made the conference and the publication of these proceedings possible:

To the William and Charlotte Parks Animal Welfare Foundation, for a substantial grant to support the conference and for matching funds toward the publication of these proceedings; and

To the following, for their support for the conference, publication of the proceedings, or both: Regina B. Frankenberg*; American Society for the Prevention of Cruelty to Animals*; Animal Protective Society of Schenectady*; Animal Welfare Institute; Bide-A-Wee Home Association*; Central New York SPCA*; Days Inn; Friends of Animals; The Fund for Animals*; The Humane Society of New York*; Humane Society of Rochester and Monroe County*; New York State Animal Control Association; Pioneers for Animal Welfare Society*; Saratoga County Animal Welfare League*; SPCA Serving Erie County*; and the Ulster County SPCA*.

The identity of anonymous benefactors will remain anonymous, in conformity with their wishes; however, the sincere appreciation of the New York State Humane Association is hereby publicly expressed.

The New York State Humane Association is also indebted to the following individuals for their unique contributions:

The many gifted speakers at the conference all donated their services.

During the conference preparations, Sondra Woodvine joined NYSHA's staff (which, until that time, comprised one full-time employee). Countless challenges were overcome because of her efficient, reliable work before, during, and after the event.

Patricia Valusek made a gratefully accepted offer to act as one of the conference moderators, a duty she carried out with proficiency and poise.

Tesa Sallenave assisted with registration and a host of details essential to the smooth operation of the conference. Diana Henley, of the Bide-A-Wee Home Association, provided similar services, as did other Bide-A-Wee volunteers whose work she co-ordinated: Margaret Borgstrand, Estelle Hollander; Marie McCrum; and Matilda Shookhoff. Tina Ashton, Jean Daniels and Linda Rydant also contributed indispensable administrative assistance during the conference.

Hilda Daily; Edward Ashton (of Animal Right Advocates of Western New York); and Ronald Scott (of Argus Archives) volunteered their services in recording the event—in photographs, audiotape, videotape, respectively.

Barbara Esmark provided guidance in the graphic design of the present volume. Steve Gulick, in addition to filling a variety of roles at the conference itself, devoted countless hours to NYSHA as its word processing and computer technician, and assisted greatly in preparing the transcript of the proceedings for publication. Robert Haiber, of The Info Devels, Inc, was exceptionally accommodating in the creation of the transcript's layout, and in converting the material to camera-ready copy.

George Fletcher, of Fordham University Press, was key to the publication of these proceedings. Because he readily understood the value of the information that would be contained in this volume, he was willing to give his strong vote of confidence in the project to those of his colleagues who had the responsibility of deciding whether or not to undertake the publication of our manuscript.

During the planning of the conference, it was apparent to the Board of Directors of NYSHA that undertaking an event of such scope would require an outreach effort. The directors, before appealing to others for funding, themselves contributed financial support.

Samantha Mullen
Public Affairs and Programs Administrator
New York State Humane Association

* Member of the New York State Humane Association

# STATE OF NEW YORK
# LEGISLATIVE RESOLUTION

Senate No. 1050

**By:** Senator Cook

Assembly No. 1304

**By:** M. of A. Hinchey
Brennan, Bush, Catapano,
Connelly, Connor,
DiNapoli, Gaffney,
Greene, Hoyt, Parola,
Pataki, Pordum
and Tonko

To support the aims of the New York State Humane Association at their fall conference, in relation to their attempt to educate the public about the critical pet overpopulation and to address the problem with humane means

**WHEREAS,** Overpopulation of dogs and cats exists throughout New York State, resulting in an enormous number of stray and feral dogs and cats; and

**WHEREAS,** Such animals frequently cause health problems in cities and suburbs, and are a hazard to wildlife and farm animals as well, in rural areas; and

**WHEREAS,** Stray and feral dogs generally suffer intensely during their lifetimes and are likely to meet painful deaths; and

**WHEREAS,** Control of strays represents a significant expense for local governments as well as a frequently overwhelming burden for animal control personnel and the staffs of animal shelters; and

**WHEREAS,** The large stray population is due to uncontrolled breeding of pet cats and dogs; and

**WHEREAS,** The New York State Humane Association, a statewide animal protective institution, organized in nineteen hundred, incorporated in nineteen hundred twenty-five, and comprised of humane groups and individuals, is planning a conference to be held this fall in New York City, to educate the public about the critical pet overpopulation issue and to focus on responsible, humane means of addressing the problem; now, therefore, be it

**RESOLVED,** That this Legislative Body supports in principle this conference on stray cats and dogs and pet overpopulation; and be it further

**RESOLVED,** That it is the sense of this Legislative Body that it should encourage all humane measures to reduce the population of cats and dogs; and be it further

**RESOLVED,** That this Legislative Body favors the surgical sterilization of mature cats and dogs and prepayment or a deposit for this surgery for immature cats and dogs, prior to their being adopted out by animal control agencies and shelters, and be it further

**RESOLVED,** That this Legislative Body pause in its deliberations to support the New York State Humane Association in their effort to educate the public about the critical pet overpopulation, and to address this problem with humane means; and be it further

**RESOLVED,** That copies of this Resolution, suitably engrossed, be transmitted to the New York State Humane Association; to the Governor of New York State and the Mayor of New York City.

**ADOPTED IN SENATE on May 12, 1987**

By order of the Senate,
Stephen F. Sloan, Secretary

**ADOPTED IN ASSEMBLY on May 26, 1987**

By order of the Assembly,
Francine M. Misasi, Clerk

# The Speakers

## NYSHA 1987 Conference
### Overpopulation of Cat and Dogs

*The information below pertains to the conference participants' positions at the time they spoke at NYSHA's conference in September, 1987. Some—whose titles are marked with asterisks—now have different positions.*

**ARTHUR BAEDER, D.V.M.**—Veterinarian practicing in Rockaway, NJ; Secretary,* New Jersey Veterinary Medical Association, which worked for the passage of New Jersey's subsidized, low-cost spay/neuter legislation

**ROBERT CASE, D.V.M.**—Veterinarian practicing in Schenectady, NY; Legislative Chairman* and Past President, New York State Veterinary Medical Society

**BARBARA CASSIDY**—Director of Animal Sheltering and Control, the Humane Society of the United States, Washington, DC; evaluates policies and procedures of animal shelters throughout the country, and conducts staff training programs

**MURRY COHEN, M.D.**—Psychiatrist practicing in New York City; Secretary-Treasurer,* Medical Research Modernization Committee, a scientists' group concerned with the welfare of experimental animals.

**ANN COTTRELL FREE**—Author and journalist; 1963 recipient of the Animal Welfare Institute's Albert Schweitzer medal; 1987 recipient of the Rachel Carson Legacy Award; instrumental in establishing and naming the Rachel Carson National Wildlife Refuge, in Maine.

**MARGARET GERAGHTY**—Executive Director, Chemung County SPCA and Humane Society, Elmira, NY; Past President of the New York State Animal Control Association.

**WOLFGANG JÖCHLE, Dr. Vet. Med.**—President, Wolfgang Jöchle Associates Inc., Denville, NJ; veterinarian recognized internationally as an authority on the reproduction of domestic animals and on companion animal population control.

**JOHN KULLBERG, Ed.D.**—President, American Society for the Prevention of Cruelty to Animals (ASPCA), an organization based in New York City that handles about 80,000 animals each year in its shelters; investigates and prosecutes cruelty cases; and promotes animal protective legislation at all levels.

**MARTIN KURTZ**—Director, Bureau of Animal Affairs, New York City Department of Health.

**LEO LIEBERMAN, D.V.M.**—Veterinarian residing in Port St. Lucie, Florida; has extensive experience in the neutering of immature animals, and has published on this subject.

**JOHN McARDLE, Ph.D.**—Anatomist, Scientific Director, New England Anti-Vivisection Society,* Boston; board member, Scientists' Group for Reform of Animal Experimentation.

**ELINOR MOLBEGOTT, ESQ.**—General Counsel and Vice-President for Legislation* at the ASPCA.

**INGRID NEWKIRK**—Co-founder and National Director of People for the Ethical Treatment of Animals (PETA), Washington, DC; former director of cruelty investigations, Washington Humane Society; and former chief of animal disease control for Washington, DC.

**TOM REGAN, Ph.D.**—Professor of Philosophy at North Carolina State University, Raleigh; the author of *The Case for Animal Rights*, and numerous other major publications on animal rights.

**ROBIN REMICK**—Executive Director of the Tompkins County SPCA,* Ithaca, New York; Past President of the New York State Animal Control Association.

**GORDON ROBINSON, V.M.D.**—Vice-President of Veterinary Services at the ASPCA, and Area Director/Region 1, American Animal Hospital Association

**DAVID SAMUELS, V.M.D.**—Veterinarian practicing in New Paltz, NY; participant in several low-cost community service spay/neuter programs

**RONALD SCOTT**—Director, Argus Archives, a New York foundation which compiles information on the protection of animals; presents and produces video documentation of humane education programs in Garrison, Tarrytown and New York City schools

**CHRISTINE STEVENS**—President, Animal Welfare Institute and Secretary, Society for Animal Protective Legislation, Washington, DC; Past President and honorary board member of NYSHA

**SHERI TRAINER**—Director, Educational Services, Bide-A-Wee Home Association* (NY City, Wantagh, Westhampton).

**ANNABELL WASHBURN**—President, Pet Adoption and Welfare Service, Inc. (P.A.W.S.), located on Martha's Vineyard, Massachusetts.

**GRETCHEN WYLER**—Vice-Chairman, The Fund for Animals; Honorary Chairman of the New York State Humane Association;* stage and television actress.

**KATHLEEN YOUNG**—Manager, Bide-A-Wee Home Association's Westhampton shelter.*

# DAY ONE

# Opening of Conference and Welcome

**DR. MARJORIE ANCHEL:** I would like to open the conference by introducing Assemblyman Morton Hillman of the 26th Assembly District—my assemblyman. Mr. Hillman has always been very helpful to us, in local animal problems as well as in connection with state legislation.

**ASSEMBLYMAN MORTON HILLMAN:** Thank you, Dr. Anchel. I would like to officially welcome the audience to this conference. I am happy to be here. I have loved animals all my life. They give us so much, that there's no way that we can't do everything possible to make sure that those animals who can't say anything to us verbally but who say everything to us in every other way they can, are taken care of.

I think what you all are doing is probably one of the most under-publicized, non-talked about, areas in life. Somehow or other, the priorities of life just don't always allow us to think of animals. I perhaps am referring to myself as a state legislator. But you all work at it full-time in your minds and in your hearts.

I, for one, am very pleased to be one of the sponsors of a bill that's in your folder, A.4968-C.[1] I will continue to do everything I can, not just to sponsor legislation, but to make everyone in the Assembly aware of what you are doing, because I do honestly believe that if everybody really knew about it, we wouldn't have the resistance that we do.

I do want to thank you for inviting me. I promise you I shall support all of your organizations and the Association as a whole for as long as I am in the Assembly, as far as legislation is concerned, and forever as a person.

And now I have a delightful job. This is a resolution[1a] introduced in the Senate by Senator Cook, and in the Assembly by Maurice Hinchey, and passed by the legislature. I am very honored to present this legislative resolution. In essence, it recognizes what you're doing here today. I would like to present this to Dr. Anchel and to Mr. Frank Rogers, representing the Association. Thank you very much.

**ANCHEL:** Thank you, Mr. Hillman. I would now like to introduce Dr. John Kullberg, President of the ASPCA.

**DR. JOHN KULLBERG:** I've been asked to introduce Mr. Martin Kurtz. I opened my packet, as all of you did today, and I saw the letter to *The New York Times* written by Lia Albo, the new National Coordinator for the Fund for Animals, published on March 5, 1987, and there is one very relevant paragraph in this letter that I think is an appropriate preface to my introduction of Mr. Marty Kurtz:

> *Overpopulation of dogs and cats is nearing disaster proportions. Every day, more are being born and more are dying needlessly. Concerned humanitarians rescue and make provisions for many of these animals, but the task is beyond their means. Serious dialogue must begin with the ASPCA and the City for an immediate resolution. It is imperative that animal care facilities be appropriated for each borough and even more that an effective spay-neuter program be established. A responsible and progressive animal care program would include 24–hour emergency rescue, humane education programs, cruelty investigations and prosecutions of dog and cock fighting, which are now embarrassingly rampant in our city.*

I would underscore that it happens to be true that the programs mentioned, and in particular, the absence of appropriate shelter and spay/neuter facilities in each borough are, to quote Lia, "embarrassingly rampant in our city." That said, the two busiest animal shelters in America are here, and the largest free spay/neuter program in America is here, and they were built and today are run by the ASPCA. The largest free spay/neuter program in America operates out of the ASPCA's two veterinary hospitals in New York City. Nonetheless, shelters and

## Opening of Conference and Welcome

spay/neuter programs are still embarrassingly limited not only in New York but all over the country.

Here in New York, you have a not-for-profit humane society undertaking animal control, an all too common phenomenon around the country. We work very closely with the City Department of Health's Bureau of Animal Affairs, the director of which is Mr. Martin Kurtz. As president of the ASPCA, I, along with Mr. Kurtz, in open contract negotiations and behind the scenes, attempt to solve the City's ongoing animal problems. Mr. Kurtz and I have to regularly face the implications of pet overpopulation more than any other individuals in this city. Our relationship has not always been what one might consider a friendly one. We have had our differences, we have had our arguments, and I think both of us at times have hung up the phone on the other. But in the several years I've come to know Mr. Kurtz, I believe he all too well does understand the animal population dilemma. But he too is caught in the inevitable web of many financial demands on the Mayor's office and the City Council and too few resources to meet those demands. And unfortunately, and unfairly, all too often, because government is what it is, the shortfall of resources often winds up on the backs of the animals. What we're about here today is to try change that—to get the dialogue going in the community, and to bring pressure on government so that the service areas in city government, including the Department of Health's Bureau of Animal Affairs and the humane societies in the community can better work together to establish ways of undermining the growing tragedy of pet overpopulation.

Mr. Kurtz is the person assigned by the Mayor to review ASPCA City contract performance, assuring City Council representatives that appropriate fiscal, personnel and other related contract matters are properly carried out. It is a very complicated endeavor, animal control, and I am delighted that Mr. Kurtz does have an opportunity to present some concerns from his perspective this morning. Mr. Kurtz.

**MARTIN KURTZ:** I'm pleased to have been asked to say a few words at the opening of New York State Humane Association's conference on pet overpopulation. As the director of New York City's Bureau of Animal Affairs for the past nine years, I have come to realize that there are literally a myriad of issues involved in the delivery of animal control services. From a review of the topics to be discussed today and tomorrow, I can

Dr. John Kullberg

see that we're going to address some key issues. Animal control systems, euthanasia, spay/neutering, humane education, the role of animal shelters, and animal legislation, to name a few. I'm particularly looking forward to the first presentation of animal control models, and we'll be available to discuss it afterwards. Thank you very much.

**ANCHEL:** Thank you, Mr. Kurtz.

*The following talk was delivered at the end of the first day of the conference, at NYSHA's Awards Banquet. It is appropriate in these printed proceedings to present it here.*

—The Editor

# HOMAGE TO BOATSWAIN AND STICKEEN

Keynote Address at the New York State Humane Association Conference on Overpopulation of Cats and Dogs.

by Christine Stevens

After the second world war, the American animal protective movement was nearly derailed by organized attack. The medical research establishment's target: homeless dogs and cats for experiment. To a considerable extent the tide has turned. No laws to compel humane society shelters to supply animals to laboratories have been passed for a quarter of a century, and several more that were passed in the forties and fifties have been repealed. This year, New York State passed a law specifically prohibiting sale of impounded animals for use in experiments.

But the underlying problem, more companion animals than willing human adopters, has scarcely begun to solve the problem in this state. There is not a differential license fee for spayed and unspayed female dogs, a simple measure to prevent unwanted litters in use for years in many parts of the country.

New York is a major target for the middle western puppy mills that churn out pups and kittens under conditions that require the constant efforts of the U. S. Department of Agriculture inspections and local anti-cruelty societies to maintain the bare minimum standards. Congress had to act specifically to prevent unweaned pups from flooding the market, the younger the more pitifully appealing to impulse buyers. Before the federal Animal Welfare Act was amended in 1970, pups of five, four, three, or even two weeks old were being shipped off in flimsy crates to the massive eastern market. The law now prohibits their transport and sale till they are eight weeks old. This reduction in cruelty is far from solving the problem however.

What are the most practical means of curbing pet populations and ensuring that those who buy or adopt companion animals recognize their responsibilities?

Humane education of a far more comprehensive and effective nature is obviously necessary, and the work of Sheila Schwartz's group deserves commendation. Every child needs guidance. Unfortunately, though, many teachers have no grasp of the subject, having been trained in the school of thought that dismisses all value judgments.

But here, too, the tide has begun to turn, and if we pursue the matter vigorously, using well-prepared and convincing teaching aids, and making clear the benefits to society as a whole, we can prevail. The very increase in sado-masochistic behavior that threatens the common good, not to mention elementary safety for the average citizen, serves to highlight the basic need for humane education.

The sadistic savagery of those who first pitted dogs against each other, who bred them for strength of jaw and aggressiveness of temperament over the centuries, and who continue to beat them and starve them and set them against kittens and other helpless small animals, against each other and against human targets, those people should be punished to the full extent of existing law and they should be denied the right to own *any* animal.

There has been a move across the nation of make the state anti-dog fighting laws felonies rather than mere misdemeanors. Prohibition of animal ownership, as provided in British legislation following serious violations of anti-cruelty laws, would help prevent recidivism.

But the basis of this grossly uncivilized behavior must be attacked, and the gentle, broad-gauged methods of humane education offers the most promising way of preventing youths from turning to violence.

## Keynote Address

The work of Stephen Kellert of Yale University's School of Forestry and Environmental Studies is doubtless well-known to most of you. In particular, his surveys of violent criminals in prison for serious crimes—assault, rape, murder—reveal a significant correlation with extreme cruelty to animals during childhood. The studies also reveal that the fathers of many of these violent criminals were habitually cruel to animals and to their sons.

But childhood cruelty has not received the serious attention it needs and deserves. Ideals of compassion, kindness, and self-restraint should be inculcated in school, and children who show sadistic tendencies should be given special attention to guide them away from such behavior and, where necessary, to protect them from brutal parents. Child abuse, like cruelty to animals, needs far more serious concern by elected officials and the bureaucracies they rule.

In the long-term struggle against indifference to the fate of animals, cultivation of respect for them is essential, and to the extent that this is not just lip service but a strong empathy built on both reading and personal experience, its effectiveness is heightened. The many splendid television programs that have brought millions of viewers close to creatures they could never have aspired to see otherwise—certainly not on such intimate terms—build empathy and respect for wild creatures in remote places, and I believe, the dogs and cats who are our especial concern at this conference benefit from the general recognition that ours is not the only interesting species on earth.

Nevertheless, the old saw, "Familiarity breeds contempt," is regrettably applicable all too often in human relations with companion animals. Despite the fact that abandoning them is a crime, countless dogs and cats are left behind at summer resorts and throughout the year when people move. The guilty parties appear to feel neither responsibility nor compunction. They fail to realize that the creature they have unfeelingly discarded is, as philosopher Tom Regan tells us, "the subject of a life."

It is the job of effective humane education programs to plant a lively conscience in every child with regard to his or her responsibilities to all companion animals that share the household, and to extend that conscience as much as possible beyond the confines of home.

World literature, well used by teachers, can inspire young students with attitudes of true respect and understanding for the animals closest to us. America's John Muir and Britain's Lord Byron have told us in their own unique ways how highly they held Stickeen and Boatswain in their esteem.

Muir called Stickeen, the dog who lived through a blinding snow storm with him and crossed a seemingly impassable crevasse in Muir's ice-axed footsteps, "the herald of a new gospel."

"I have known many dogs," wrote Muir, "but to none do I owe as much as to Stickeen. At first the least promising and least known of my dog-friends, he suddenly became the best known of them all. Our storm-battle for life brought him to light, and through him as through a window I have ever since been looking with deeper sympathy to all my fellow mortals."

Byron's epitaph for the Newfoundland dog with whom he is portrayed in the life-size statue near London's Marble Arch is a classic:

> *Near this spot*
> *Are deposited the remains of one*
> *Who possessed beauty without vanity,*
> *Strength without insolence,*
> *Courage without ferocity.*
> *And all the virtues of man without his vices.*
>
> *This praise, which would be unmeaning flattery*
> *If inscribed over human ashes,*
> *Is but just tribute to the memory of*
> *Boatswain, a dog.*

Montaigne perspicaciously analyzed the mysterious knowingness of cats when he graciously acknowledged that he couldn't be sure when playing with his cat whether, in fact, in her opinion, she was playing with him.

Ivan Turgenev, Mark Twain, Colette, Stephen Crane, Jean Giraudoux, have shown us how we should think of our relationships with dogs and cats and with what horror we should view cruelty to them. If every high school student were encouraged to read these authors, perhaps in an anthology, some of the difficulties of applying humane education principles to the studies of teenagers might be solved.

But recognizing that education alone cannot address the many problems of pet overpopulation and stray animals, we must also turn to law and science for the contributions they are best suited to make.

Urgently needed is a readily administered substance to provide permanent or at least fairly long-term sterility. If such a substance were available, it could be widely administered, using as a model the rabies clinics at which veterinarians volunteer their time to inject vaccine paid for by the city. Many dog and cat owners who are reluctant to pay the cost of surgical neutering of animals would bring their animals to free clinics for a shot. Mobile clinics would ensure that no one would have to travel long distances in order to participate.

Despite the existence of numerous drugs and biologicals developed for human birth control, none have been fully researched for use on dogs and cats. This should be a high priority for animal advocates. If this conference adopts resolutions, I hope this goal will be at the top of the list. In this age of technology, it is inexcusable that no practical method of mass fertility control for companion animals has been developed—and this despite the fact that millions of dollars are spent in big cities every year to capture and kill unwanted stray or feral dogs and cats. On the selfish grounds of rational economy alone, adequate funds should be allowed to the development of technology to make chemical birth control a reality.

Animal control based solely on continued round-ups of homeless animals is economically unsound and should be recognized as such by mayors and city managers. They could pool resources for the necessary research and development and save themselves millions of dollars. Best of all, the distressing round-ups which, by their very nature, are inhumane, could be enormously reduced in scope if not wholly eliminated.

Until oral or injectable birth control is available for use, we much look to legislation to ameliorate the current intolerable situation.

I've already mentioned the need to charge high license fees for unspayed females, low fees for spayed as an incentive to pet owners to invest in the necessary surgery.

Another important legislative approach is to ticket and fine owners of dogs who violate the leash laws. Why should dogs pay with their lives for the carelessness, indifference, or laziness of an owner who gets off Scot free? Following a dog home is less expensive, less cruel, and less dangerous for the dog catcher than is capture and killing of the animals. Owners should have to pay a pet straying ticket in the same way as they are obliged to pay a traffic ticket. That is the civilized way to enforce the license and leash laws. Combined with incentives for spaying, it could lead to a major reduction of feral populations.

Until chemical methods are fully developed and regularly used, surgical neutering should be subsidized by local governments and animal welfare charities, making sterilization readily affordable by all.

With respect to cats, the Universities Federation for Animal Welfare has developed a remarkable system for feral cat colony control. The feeding of stray cats by kindly individuals is a well-established phenomenon throughout the civilized world. It is immortalized in one of Giraudoux's most famous plays, "The Madwoman of Chaillot," and tourists have long been astounded by the spaghetti-eating hordes of cats at the forum in Rome. Yet, till very recently, no rational approach had been made to the prevention of disease and excessive growth of such colonies. Following a careful study and successful pilot projects, UFAW has made a documentary video with accompanying literature demonstrating the system so that it can be used wherever people of good will decide to make the effort.

The cats are captured in baited box traps and examined by a veterinarian. Those young enough to adapt to home life are placed in homes, and those judged to be incurably ill or injured are euthanized. The others are given any necessary treatment, vaccinated, spayed, marked, and returned to the colony. The local cat feeders continue to feed them as before. This system allows limited numbers of cats, often welcomed by factories, hotels, hospitals, and other institutions as a defense against unwanted rodents, to remain in residence without expanding excessively. The UFAW video shows an attractive small shelter and resting place for cats who live on the grounds of a handsome hotel in Kenya.

Recently the Associated Press carried a story about a worker who had fed the cats at a big airplane manufacturing plant for many years, but the cats had multiplied to such an extent that the company had called in an exterminator and fired their valued employee. The UFAW system could have prevented all these regrettable occurrences. Indeed, I have written to draw it to the attention of the plant manager in hopes of reversing the situation.

It was encouraging to read the Legislative Resolution adopted in May by the New York State Senate and Assembly in support of this conference. Its recommendation of prepayment deposits for sterilization of pups and kittens adopted from shelters should be standard practice and should be followed up in every case to ensure compliance.

Perhaps the legislature that passed this resolution might be persuaded to place greater restrictions on importation of the "puppy mill" animals that contribute to the shameful "throw-away" attitude towards companion animals because they are so aggressively merchandised to anyone who will pay the price at which they are sold. And, of course, there is no obligation to have them spayed.

I hope this conference will lead to decisive action by legislators, educators, mayors, and city planners and the scientific institutions that can provide the expertise to make mass sterilization feasible for our dog and cat friends—in John Muir's phrase, our "horizontal brothers." It's time, indeed long past time, that the human race should take seriously our huge debt to them.

# ANIMAL CONTROL:
# HOW IS IT BEST HANDLED?

**ANCHEL:** Our first topic will be, "Animal control: How is it best handled? Rural, urban and metropolitan approaches." Samantha Mullen will introduce the speakers.

**SAMANTHA MULLEN:** This is going to be a panel discussion. Margaret Geraghty, Robin Remick and Gretchen Wyler will be speaking.

Margaret Geraghty is president of the New York State Animal Control Association. She has been a member of that organization since 1985, and two years prior to that, she became director of the Chemung County SPCA in Elmira. That shelter, as well as the shelter of which Robin Remick is director—the Tompkins County SPCA in Ithaca—are generally recognized by people in the animal welfare field as two of New York State's most efficiently run animal shelters, and we're proud to have the directors of those two shelters as speakers today.

Both Robin and Maggie are seasoned professionals in the field of animal welfare. They're striving continuously to gain new knowledge about their field and to impart it to the members of the New York State Animal Control Association. I've introduced them simultaneously because they work so closely together, and because they represent the same kinds of ideals. We are very grateful to them for the work that they do on a daily basis in operating animal shelters. It's one of the most difficult and challenging jobs there is, and they both do it superbly.

To my left, Gretchen Wyler. I'm very pleased to introduce this wonderful champion of animal rights who has been working in the field while maintaining another challenging career, that of dancer, singer, and actress. As an animal activist, she's known to all of us as somebody who's willing to give of her time and her talent, virtually whenever we've asked her. She serves currently on the board of directors of the Fund for Animals, of which she is Vice Chairman and, of course, as you know, she's the honorary chairman of our own organization, NYSHA. She is the Vice President of Beauty Without Cruelty, American Fund for Alternatives to Animals in Research, and she also serves on the Mountain Lion Preservation Foundation and three political action committees.

Margaret Geraghty will speak first.

**MARGARET GERAGHTY:** I'm delighted to be here, and one observation I made as we came up here is, "We don't look like the dog catcher, do we?" Animal control has changed. Animal control is better, more professional. I'm talking about world animal control. I'm the director of the Chemung County Humane Society, Elmira, New York. We contract with nine other towns in Chemung County for animal control purposes. We do have a city, we have some suburbia, but mainly it is country. Attitudes in the country do vary from attitudes in the city or suburbia, but on the whole, the specific philosophy of animal control is the same, no matter where you are.

I believe that the first thing the animal control officer, who is humane, must do is to respond to immediate problems caused by or to the stray animal. I also think that we must assist the citizen with any of the problems that are caused by the stray animals—dogs in garbage, cats in gardens—helping them to solve those problems. We do have to address public health and safety concerns, rabies concerns, dog bites, motor vehicle accidents caused by strays. Also, we need to prosecute—prosecute the irresponsible animal owners. I do look upon that as a negative form of education, and I do believe that education is the most important thing that the animal control officer can do, but if you can't convince the person, you may have to prosecute. Also, we need to do as much as we can in promoting the end to animal overpopulation. That's why we're all here. Also, we're there to provide medical assistance and oversee the proper care of the animals that we have. That, basically, to me is not a problem of urban, suburban and rural.

In the country, you're going to see some attitudes that may surprise you. The first thing that the rural animal control officer has got to do is know his area. He's got to know Park Road from Jackson Road. But besides the physical, he has to be aware of the differences of the people, the attitudes that they have. In the rural area, most people are used to seeing their animals loose. Boundaries are not as precise as they are in the city. Also, in the rural area, the people are more in tune with animals being born, living and dying as a commonplace occurrence. Domestic pets are not seen just as providers of comfort and love, they're also looked at as workers. They're out there doing their jobs. The farm cats are mousing, the dogs are herding cattle or maybe herding the kids.

Licensing is a problem in rural areas because licensing is not always seen as a necessity. In an economically poor area, that's one of the last things that they're going to pay for. Then you have the town boards, who feel enumerations—dog censuses—are not mandated, so therefore, that's an unnecessary economic burden, to perform an enumeration. Then there's the common myth that the farm dog does not have to be licensed. I don't know where that started, but it takes a lot of education to let people know that, whether they're on a farm or not, they need to be licensed.

Another thing I noted is the rural citizen seems to have closer ties with his town officials. Usually, they're related or they're neighbors, and you can be guaranteed if the supervisor's brother has a problem with loose pigs in the neighborhood, that you're going to be pig control, whether you are by contract or not. I think these are the predominant attitudes that rural people have, that you have to address.

The complaint level seems to me to be very low in the rural area, because most of the people feel that they can deal with the problems themselves without the assistance of an animal control officer. I will, however, tell you a story of what happened.

Stray dogs running loose don't mean a lot to rural people. In 1982, our shelter got a phone call from a farmer about some pups that were abandoned in a small town, Baldwin. There are probably 800 or 900 people in Baldwin. We've contracted with them since 1980, and we have an average of 30 calls a year from the town of Baldwin. The call came in to pick up the stray pups. By the time the officer got there, the pups were gone. No one had any idea where they were. One year later, we started receiving calls from the people in Baldwin that livestock were being killed. It took us two weeks of setting humane traps before we were able to trap two feral dogs. The third one ended up being shot by a farmer. Those are the three pups that we didn't pick up in 1982. One year later, never having known any human kindness, any human touch, they were out killing livestock. That's a perfect example of the problems that we have with strays.

The secluded nature of life in the country is perfect for the cultivation of one of animal welfare's prime concerns: animal blood sports—dog fighting, cock fighting. The rural officer can see some indication if he sees excess traffic at odd hours around isolated buildings, wounds on an animal's body. The best method, though, is the informant, who is very rare. The people who are involved in dog fighting or cock fighting are aware of the laws. They're aware that what they're doing is illegal, and they are underground. I always tell my officers if there are any indications, we're going to contact national and state organizations to assist us, rather than handle it on our own. The other illegal activities, such as drugs, illegal weapons, that could possibly be going on, would interest the local police agencies also.

There's always a question of cruelty to livestock. You and I both know that the horse that's out there on the hillside in the blizzard is out there because he wants to be, but there are a lot of people who don't realize that, and the rural animal control officer has to know what actually is cruelty and what isn't. Know your county cooperative extension, know the farm grange people, and get on a first name basis with the large animal vet who does go to these farms daily and can tell you which animals are being neglected, and who needs your assistance.

Nuisance wildlife is a recurring problem. Most rural people take care of their problems themselves. Wildlife is under the direct auspices of the Department of Environmental Conservation. But many animal control officers do have it in their contract to assist with sick, injured or vicious wildlife. In order to do that, we need a nuisance wildlife permit. To get that permit, you need to take the trapper training course, which many of us find not to our liking. We do not want to be involved in direct trapping. The New York State Animal Control Association yearly puts on a program under the auspices of DEC. It's a modified trapper training course, and the emphasis is put on humane live trapping, humane treatment of animals and disease identification and control.

New York State [Agriculture and Markets] law, Article 7, mandates that the fiscal responsibility for dog control is going to be on towns. Therefore, it is important for that rural animal

## Animal Control: How Is It Best Handled?

control officer to be able to justify his budget and to be able to put in enough to show that he will able to provide humane dog control. His effectiveness and the budget really will set the tone for the animal control for that next year.

To recap, I say the philosophies are the same, rural or urban. The attitudes that you're going to find will be different. You have to work through those attitudes. Find the key person who believes in animal control in the rural area, who will follow what you are saying, who will be the law-abiding citizen, use that key person to promote humane animal control. Also, the number one tip, I think, in rural animal control, is know your town clerk. He or she is the one you should work with. He or she has their finger on the pulse of the community. They know everybody. They know the politicians. If you can work with the town clerk, you will be able to let him or her know why licenses and enumerations are needed, and how they will benefit the town: better identification of the dog population and more monies to conduct a humane animal control program.

New York State Animal Control Association is addressing the fact that there are very few schools for the animal control officer. To date, there are none for preparation. Once in the field, there are national humane and animal control organizations that do provide schooling. New York State Animal Control Association is doing likewise. We present workshops for the animal control officer on topics that they recommend.

A recent membership survey showed that, of the top three concerns of the animal control officer in New York State, number one was mandatory qualifications for the position of Animal Control Officer.[2] In order to control animals humanely, the people hired to do it should at least meet equal minimum standards. Two, mandatory on the job training for the ACO positions, tied with improved image of the position. I feel that if animal control officers take advantage of training programs and conferences, they'll attain the competence that's needed and take pride in the work that they're doing.

Better state enforcement was the third priority. The state is needed to assist us, when we find out that there are unacceptable animal control facilities. We need to be able to turn to the state knowing that they will back us one hundred percent, closing the inadequate animal shelters. That way, we'll know that they have a firm belief in the laws that they have made.

Margaret Geraghty

Animal control and animal welfare walk hand in hand. You can't have good animal control without having concerns for the animals you're caring for, and you can't have good animal welfare unless you set up controls and follow them. I do feel that with cooperation, working together, we can achieve that for the animals that we're here for today.

**MULLEN:** Thank you, Maggie.

## Animal Control: How Is It Best Handled?

**MULLEN:** Robin Remick is now going to address the problems inherent in animal control in urban areas.

**ROBIN REMICK:** I'm from Ithaca, New York. It's not all that urban. It's a small city where students outnumber the natives, and every dog that comes into the animal shelters has a red bandanna around its neck—no tags. There are inherent problems in communities where the people that live there don't plan to live there for long, and Ithaca is a very transient community.[3] People who don't have to live next door to each other for very long don't mind, maybe, offending people with their pet problems, and there is a lack of community mindfulness and consideration for each other's property, and animal problems are rampant.

There are some things that can be done and that you might want to look at in your own communities, especially in tourist towns and college towns, places like that.

Apartment managers need to establish clear, written pet policies and to enforce them. These policies should be based on community respect for each tenant's needs, both pet-owners and non pet-owners. And this policy should definitely include a no litter clause.

Animal overpopulation is a real problem. Local animal control laws should reflect the needs of that community, and they should perhaps be revised to include penalties against apartment owners if the tenant's pet is allowed to become a nuisance. If the landlord doesn't enforce his own policies, but then he expects the town animal control officer to come in and do his job for him, it's very difficult. It really should begin right with the apartment owner. Leash laws, pooper-scooper laws, barking dog ordinances, etc. When I go to classrooms and speak to children, I always tell them, "My neighbor doesn't like my dog as much as I do." And that's from where we start.

We need to discourage the feeding of stray cats and dogs, allowing them to linger in the streets. This hit and miss method only leads to trouble later on—hit by cars, animal disease, overpopulation and all the rest. Call the animal control officer or humane society immediately. Someone might even be missing that pet.

There seem to be seasonal trends, especially in transient communities. I know in Ithaca, in the fall, students are coming in by the carloads, wanting to adopt animals. And then in May, animals get booted out and will probably be left on campus. Every summer, we get lots of stray animals. And it's the same in tourist towns, I'm told.[3]

Set up a booth on campus to promote responsible pet care. Try to become part of student orientation. Have information available about dog obedience and neutering. Supply people with copies of the local ordinance. This should be ongoing exposure, no matter where you live, by the animal control officer or the humane society.

Look to who enforces animal control in your community. Are they a part of the problem or are they a part of the solution? I work at an animal shelter, and many times, I feel that I represent some of our worst offenders.

New York State law requires each municipality that issues a dog license to hire someone to enforce that state dog law. This might be a private individual, or it might be the SPCA or local humane society. Whoever is responsible in your community, we need to know that they're doing all that they should do. The public may expect more from the humane society or the county SPCA because they figure that they're a big agency. I know in my own town, people are very accepting of the animal control work that's done because the woman who's doing it is doing the best she can with what she has. That's just not good enough.

New York State needs to enforce its own laws. They say that animal shelters need to be in compliance, they need to store their food in a dry place—there's a whole list of things that New York State has inspectors for—inspectors who come in and inspect their facilities. But when someone is found to be in violation, New York State tells us that they don't have anyone to enforce those laws, and so the people who are in violation are allowed to keep on being in violation. And because of that, we just spin our wheels in the community.

The SPCA cares for all animals. It has a broader focus and different goals than the person who's hired just to care for dogs. And so, as a person in the humane society and the SPCA, I solicit dog control contracts, more because I think animal control goes hand in hand with humaneness, and I also feel that I'd rather be doing it myself than have somebody else not doing it adequately. I think it's inexcusable to allow inhumane animal control to go on in a township.

## Animal Control: How Is It Best Handled?

Check out who does your animal control. Do they provide a 24-hour emergency service with adequate equipment? Can they really respond? Do they have trained staff, radios, vans, whatever it takes? Do they have cooperation with the local veterinary association to provide for routine inoculations of animals that are brought into the shelter? When an animal is brought into the shelter, before it's put in with the general population, that animal should be inoculated, possibly wormed. There should be emergency care for animals hit by cars. We're fortunate in Ithaca to have Cornell University Veterinary School provide this service for us at no cost, and I realize that in many places, budgets restrain this. But again, for people who care, show that you care by going to your town board meetings or city council or whatever, let them know that it's a priority with you that money is allocated for this. No one else is going to do it. It's obviously not a priority for the town boards.

Does your animal control officer have a lost and found clearing house, so to speak? Do they take reports regularly and review them regularly?

Are humane education programs offered?

When someone redeems their dog, if their dog was picked up at large and they go to the animal control officer and they happen to be my neighbor and it's a small community, and I say: "You don't really have to pay this time. However, New York State law says that before I can give it back to you, you have to license it, but will you promise me that you'll do it, because I know it needs a rabies shot done first, and I don't have a veterinarian here to do it, so could you take care of all that?" It just doesn't work, and there needs to be strict enforcement to discourage it from happening again. Get an incentive for people to obey, whatever it takes. Use the few laws that there are in your favor to keep those animals off the streets.

My biggest push of all is adoption policies. Counsel prospective pet owners regarding the specific responsibilities they are assuming. Supply them with information regarding the local pet law. Consider their lifestyle and the pet's temperament. Don't just hand out animals and look for high adoption rates. I would rather adopt out ten animals in a year and know that those pets were not going to reproduce or be ill treated, know that they were going to be able to live permanent lives in the homes, than to adopt out 20 and have those animals go out with no neutering, reproduce and cause untold problems.[4] So don't look for numbers in your adoption program, look for the quality of the life that the pet's going to lead. Refer to the list of dog obedience instructors. Use the contacts that you have in your town, kennel club, dog obedience, veterinarians—whoever it is—and offer to take the animal back if the animal doesn't work out in the home, so that it's not abandoned. Animal shelters and animal control facilities have a responsibility to the animal forever, really, and so I think that that's a real priority.

I think that in New York State it *is* a problem, because the law mandates animal control to be offered in a community, but there's really not enough money for it. It's not a priority with anyone.

One last area I wanted to talk about was wildlife control in the city—not in a big city. Many times, people will see an animal in their yard, a raccoon or a wild animal, and they automatically assume it has rabies because if it were living a normal life, it would be in a forest. And so animal shelters need to provide information that will educate people about how they might make their yard or their home less attractive and less accessible to wildlife, if they feel they can't live in harmony with wildlife—those creatures of the night that dump our garbage cans.

Educate them to seal off entries, screen the top of the chimney so the raccoons can't make a nest in it. Have a list of wildlife control agents available to these people if they do need to have the animal live-trapped at their home and removed. And again, become involved yourself. It's not a priority for the town boards, because it's not mandated that a certain amount of money be allocated to the animal control programs. Because of that, sometimes the most humane facility is not hired for the job: they always take the cheaper route. I would encourage everyone to become a part of the program locally as well on the state level. I think the local level is what really can begin to make a difference. And do what you can.

**MULLEN:** Thank you, Robin.

## Animal Control: How Is It Best Handled?

**MULLEN:** Gretchen Wyler is going to address animal control from the metropolitan perspective.

**GRETCHEN WYLER:** Thank you. I'm very, very glad to be here. I still say to people: "I'm a New Yorker who recently moved to Los Angeles." But I've been gone for nine years from this city. I was delighted when New York State Humane Association asked me to come and talk about the subject of animal control, which was my first love. I'm very glad to share the panel with these two speakers, Robin and Margaret. I was delighted with everything they said. We have such a different perspective now about animal care and control than when I started twenty years ago—it's become a much more sophisticated kind of endeavor.

I will soon celebrate twenty-one years as an activist in this movement. It all started when I discovered the local dog pound in Warwick, New York. I was not shopping for a cause that day. I don't think anybody ever does shop for a cause. If you do, you'd probably pick something chic, like a disease: you know, fight against a disease—that's a chic thing to do. Certainly, animal rights, animal welfare, is not chic, so I wasn't shopping for it, but I was a breeder in those days of Great Dane show dogs. Someone stopped me in the supermarket and asked: "You love dogs, don't you?", and I said: "Yes, I love my own," and she said: "Have you seen our dog pound?" and I said: "No, I didn't know we had one," which is not an unusual thing to say. Most people don't know if, and certainly don't know where the pound is. But I went that very day, because there was something in the look of this lady's eyes. I found a very medieval dungeon where the rats were killing the puppies, where once a week they came around from New Jersey with a truck and took the animals off for research, and I saw this in the village dump amidst burning garbage, and vowed to do something about it. So that happened twenty-one years ago in December, and I had no idea what it would lead to—an entire change of my life.

I'm very active right now in California animal rights politics as Vice-Chairwoman of the Fund for Animals, and also, I sit on twelve other boards of directors. But even though I'm doing all those other things, as I told you, my heart is in animal control.

However, I don't work in animal control at all in Los Angeles, because it's so beautifully taken care of, they don't need me. I was very involved and cared very much about animal control in New York City. When I wanted to get the little shelter built in Warwick, New York, as a result of that grim day in December '66, I went to the politicians and rallied the public, because ultimately, the only way you can get anything done is if you stir the public. The greatest motivating force in government is people, and you have to rally the people. You can't win just because you're moral and good and right. You have to get people rallied, because ultimately, the politicians will respond to people, that's what keeps them in their jobs. I don't think Warwick, New York, is any different from New York City, is any different from Los Angeles City, or Ithaca or wherever. To get something done, you have to make it happen.

When I came on the board of the ASPCA in 1972, I was the first *woman* ever on its board, and the first *person* ever to be dropped. Yes, I was dropped, because I sued my fellow board members for corporate waste and indifference to animal suffering. Promptly, I was dropped. But it was an interesting experience, and granted, the ASPCA at that time had shelters in the five boroughs. I was not a one-woman show at all. It was shared by the people of this city. We had a great constituency built, trying to reform the ASPCA. At one time, we had 93 volunteers in the shelters, in I think, a very good program, just assisting in adoptions. We settled the case out of court in 1975. I believe now the ASPCA has a heartbeat, and that's what we wanted. John Kullberg and I worked together in those grim days, and when John went on as Acting Director, we hoped he would stay.

I also happen to know from having been on the board in those days, that the ASPCA does not want to be the dog pound, and therefore, before I spoke about this subject, I said: "Please check with John Kullberg—because I don't want to rave about Los Angeles City animal control if John would resent that." and I was assured that John Kullberg would not resent me complimenting LA City.

I'm very biased. I was in *this* city in animal rights for 12 years. I've been *out there* for nine years. So, obviously, I know both situations very well, and I don't happen to think there's a great difference in the two. People are people. Wherever you have people, you have pets. Wherever you have pets, you have animal problems. And I think New York City *should* have exactly what LA City has. I don't think you should settle for less. I think you've settled for less in this city. I know you're

## Animal Control: How is it Best Handled?

not all from New York City so, understand when I say: "you," I mean New York City, because it bothers me that New York City is not the "state of the art." When it comes to animal control, I think New York City *should* be state of the art. There's no reason not to be.

Let me give you some statistics first, because I think that's important—stats on LA City. By the way, LA City is one of eighty incorporated cities in LA County, so they are not to be confused. It's just a city within Los Angeles County: 465 square miles, 4,400,000 people. It has six shelters, wonderful shelters, which have 200 paid employees.

It has 120 reserve officers. I think this is one of the most marvelous things they do. Every year, about twenty-five to forty go through a very rigorous, extensive training, and then have to pass a very difficult test, and then they're made a reserve officer, and for that, they give 18 hours a month for no pay, working with LA City animal care and control. It's a wonderful program, and I certainly think any city can have it. People always say: "I'd love to help," and very often, there is no program for them. I commend LA City for having put together such a program, because I can't tell you what 120 men and women out in the field trained to be reserve officers in animal control means to any city. I think it's one of the best things they do.

They also have something called VISA volunteers, Volunteers in Service to Animals, similar to the program that we had at the ASPCA in the early '70s. These are mostly women (there are some men) in all of their shelters, assisting in adoptions, having the time to spend, the time to talk to people who have come in to adopt animals. They have 51 trucks in LA City animal care and control, out on the road, patrolling, picking up and I think doing a wonderful job. They're open six days a week from eight in the morning until seven at night. They are very accessible to people, not only to those looking to find their animal, but also to those looking to adopt an animal. They handle close to 100,000 animals a year.

Even though we're saying the surplus animal problem is rising, I don't know the stats, I would say it's not rising. In LA City we are down, probably covering around 85,000 animals this year. Between 5,000 and 10,000 of those were picked up in our patrol trucks, because that is a very active service we provide the community. About 20,000 were calls to pick up, which is, of course, a service provided in the city. The rest were surrendered animals. Fifteen percent adoption rate, on or about, but I think I agree with Robin and Margaret about the kinds of adoptions we want. The days are over when we can proudly say: "We adopt out 90 percent of our animals." I remember those days, and I'm sure many of you in this room do. That was the most important thing we could say, and that no longer is. We talk about *good* adoptions now.[4] About 60 percent are euthanized, using intravenous injections by veterinarians and/or top qualified, certified animal health technicians, who are licensed in the state of California.

I think the most important thing an animal shelter can do is try to return animals to owners. I don't know how it is at the ASPCA this year, but I know that was a main concern of mine. Every animal that comes in should be cross-filed. I think that should be a main function of a shelter, and should never be relegated to a volunteer program. Wow, if we're doing nothing else, aren't we trying to serve people too, people that have lost their animals? I know that the first thing I ask when I go to a shelter, is: "Tell me about your RTO, your "Return To Owner"—how do you handle that?" And it's shocking very often to see a bulletin board filled with yellow papers that have been stuck up there for years, that no one even looks at, and they don't have a top-paid person to carefully file cards on every animal found, and take down the description of every lost animal and try to match it with the descriptions of those found. In LA City, every animal in each shelter is computerized, to try to help people find their lost animals.[5]

Last year, 225,000 licenses were sold. The Director of Animal Regulation, Bob Rush, contracts it out to the private sector, to help sell licenses. From that contract they netted $450,000, which represented over 100,000 licenses. They're progressive. It costs $6.5 million to run LA City animal control, 48 percent is offset by the income. Also, Bob says, he thinks they do such a grand job that in the last couple of years, they've gotten over $2 million in bequests! The city is proud of the animal control, and as a result, people leave money to it. It is considered an important agency in LA City. It's not called "Animal Control." It's called "Department of Animal *Care* and Control."

There's somehow a feeling within the humane community that if it's city run, it'll be bad. I disagree totally. We are so quick to think if the city's doing it, it's bad. Why should it be bad? Why should it be bad if New York City did it? There's a

## Animal Control: How Is It Best Handled?

heartbeat in every person, if you can find the right person, and I think New York City can.

I'll tell you another thing they have that I think is outstanding, and I don't know of any similar one around the country: they have the Department of Animal Regulation Commissioners. I wish even every small city had one. It started in 1927. The mayor appoints (obviously, he is helped in his appointments, because this is a very important commission) he appoints five people. City Council then confirms their appointments, and they sit as Commissioners, non-paid. They meet twice a month at one o'clock in official headquarters next to the City Hall. All decisions regarding animal care and control are made by these appointed commissioners. If you have a problem and you live in LA City (and I go to those meetings a lot, because the Fund for Animals works on some of their spay/neuter programs) if you go there, and say: "I don't like the way you people are doing something, I don't like the fact that I called you at 4:00 in the morning, and you didn't come." Anything little to the biggest, that's where you go, and that's where you voice it. They are accountable to you. They listen to you, five people appointed by the mayor. And, granted, even though they are political appointments, it's always our people, and I say "our people," people with hearts, people that care. It's not a lot of strange people like the Fish & Game Commission, which is appointed by the governor, that are all hunters! These are all people interested in animal care and control. I think the idea of a Department of Animal Control Commission should be mandatory if New York City could pull off what I hope it can with New York City animal control.

In 1952, LA City started selling for research. In 1981, May, it was ended by the City Council vote. Fund for Animals was one of the leaders in the repeal fight. Here again, great cooperation within the city. That was what I did for a living for about six months. But we won, and guess who was our most important person there? Bob Rush, the head of animal control, who was activist, who was aggressive, who was a great speaker. I mean, he's City-paid, he's a City person, and the reason I keep saying that is I'm afraid many people in New York City are afraid if it's "City " it'll be bad. Why? Perhaps if we had this open to questions right now, you'd tell me why. Well, I'd disagree with you still. I think to negate that would be very, very wrong.

Spay/neuter in LA City: We have six shelters, we have six

spay/neuter clinics. We do 9,000 surgeries a year at the right price. In three and a half years, LA City Animal Care & Control has spent $250,000 to augment neutering programs within the City. We do free spay and neuters for people over 65, we do free spay and neuters for family incomes of $11,000 or less. Clearly, LA City cares about pet overpopulation. In 1971, 11 percent of all licenses sold, 11 percent of the animals were altered. In '84-'85 season, 39 percent were altered, because licenses were half-price if your animal was altered. Clearly, it's making an impact.

The Have-A-Heart program, which the Fund for Animals administered for three years, is a very interesting program. I would like to touch on it: Have A Heart, here again, came about from this five-member Department of Animal Regulation commissioners. By the way, the commission rotates, of course. They serve at the mayor's will. It rotates from time to time. Someone wants to leave, he appoints someone else. We have anti-vivisectionists, we have animal rights lawyers, we have animal control people, all ranges on that commission, a lot of great voices can be heard, and I think real good ears to listen to them. The spay/neuter Have-A-Heart program came about because they require a deposit when you adopt animals at the shelter, and for years, this money went into the general fund. That's bad. Now it helps fund neuter operations.

In New York, the population is very transient. People move constantly. You lose track, you don't follow up, and all of a sudden, maybe there's $70,000 sitting there. Why not let it translate into spay/neuter? That's what it was supposed to do. So if it didn't help the special animals it was contracted for, why not let it do the job for animals that it wasn't contracted for? In Los Angeles, for three years the Fund for Animals administered the program. The City gave us the money. The Fund for Animals paid for the *administration* of it. It was a neat program. It happened one month a year, just one month, because we wanted good cooperation of the veterinary community. We had eighty veterinary hospitals work with us. They all signed up for one month. I guess we didn't threaten them, because it was only going to be one month free, to anybody that called in. Three thousand were done in one month—three thousand spay/neuter operations by eighty-one vets in the City of Los Angeles through the Have-A-Heart program. This year, like every year, we spent about $50,000 to $60,000 on that. I think that's a neat program, and I wanted to bring it to your attention, whether or not you would want to do it in your town. I wanted to tell you about it, because it came out of LA City animal control, not a humane society.

We have wonderful humane society shelters in LA. But they can't accept stray animals. We've got twenty-seven facilities in the entire county, which is a lot. That's six city and six county and then about twelve others, because like Bide-A-Wee, there are many that do not destroy animals. You know, "We never kill an animal" kind of philosophy. So they start an animal shelter, and that's what they do: they never kill an animal. But, as a result, of course, they can only take very few, because their cages are always full. But the point I'm making is that there are a lot of shelters in LA County, but the LA City shelter to me, is as humane as any of the humane society shelters. That's why I wanted to bring that to your attention. It could happen here in New York City.

They also are opening a new shelter in a couple of months, and I cannot tell you the excitement I felt about it when I saw it. For all cats and puppies, Bob Rush has devised a cage system where it looks like the animals are free. It's all glass, and you look in glass windows, and then the cages have no back. They're open in the front, open in the back, so there's air going through. They have a humane education room. They work with the unified school district. Every day that they're open, there will be a bus of children coming to tour the new facility, to tour the LA City's newest shelter, young children that will be walked through, explained animal control problems, explained animal care. So, I think that's obviously, most commendable. I'm very excited about the new shelter opening. It will be a festive event. I guarantee you, the City will be proud, the mayor will be proud, the City Council will be proud, because they can honestly say that a part of the good city called LA is good animal care and control.

Those are the stats, obviously very "editorially" presented by me. I am very biased. I think it's "state of the art." I resent the fact that New York City isn't. I don't know what's going on in the City. I know the ASPCA does not want to be the dog pound. I don't think they should be the dog pound. I think it should be City run. I think it should be a recognized City agency accountable to the people of the City. I don't think it should be the private sector, somebody you contract with. I think it should be the City, and I think there's a fear in this humane community in this town (perhaps I'm wrong, I've been gone awhile) that if it's "City," it will be bad. And that's wrong.

You know what I'd do? Here's what I'd do. (Don't worry, I don't live here, and I won't do it.) If I were here, I probably would take a year out of my life, perhaps someone in this room could. I was thinking, coming on the plane last night, "You can't wish it to happen, you can't will it to happen, but you could make it happen." And how make it happen? The people have to want it. I suggest to you that you go for a referendum in this city. I would start immediately to go for a referendum. I would get it on the ballot, if not next year, in two years or in three years. I would hire a PR firm, I would collect the money to make it a popular issue in this city. You can't tell me that people would fail to get signatures and then get it on the ballot, if it were properly presented: "We are a city of X million people with X million animals. We have two shelters where we should have five; it's not accountable to the City; there are no trucks out picking up; if late at night, your animal gets hit, there is nowhere to go."

You could sell that and get it on a referendum, and I think you would overwhelmingly win the referendum if you do it combined with a top PR firm: I mean bucks. I don't mean out of your basement mimeographing. I would hire a top PR firm for the last three months before the vote is taken, and I'd have it on every pillar and post. I don't think you'd lose this one! I really don't think you would, if you did it carefully. I don't know who you are, but I would hope that there's somebody sitting here today that might want to do that for a living, decide that's what they want to do, and get it done. You could also sell it to the people. If they say: "How much money?" Look at the City tax that is realized from pet food sales. Look at the millions of dollars they get in City tax, so it's not that they're taking money from their money—taking money from pet food and putting it in. Maybe it costs the City. I don't care how much money. That's not our problem. Our problem is to get the people to say: "Do it." And you have a right to have the people say: "Do it." The City is not doing it. Why? Because the ASPCA is there, and so, obviously, they love the fact that the ASPCA will take that problem. Don't make it so easy for the City. That's my suggestion. I think the climate is right—right now. I think you can rally the people to do it, and I think a referendum is the only route. Since I got involved in animal care and control in this town in 1970, when I met Cleveland Amory at a party, and he said, "Would you like to join me? We're suing the ASPCA." That's what started me. I know this city, and I know what the ASPCA does and doesn't

## Animal Control: How Is It Best Handled?

want to do. I think, as John and I talked, it's a fine humane society, the ASPCA. It doesn't want to be the dog pound. It's not geared to be the dog pound. I don't think it *should* be the dog pound. I think that should be the problem of New York City. I think Bob Rush should be brought in. I think he should be brought in as a consultant and paid to set up the same system, and you all could do it, and I don't think you should resent the fact that Bob would come in. I don't think New York City should be threatened by something LA City does. It's called "why re-invent the wheel?" So, I wish you well. I hope someone is hearing me. Otherwise, it's been fun for me to think about not having to get those phone calls from many people in this room, saying: "Gretchen, we don't have animal control in this city. Do you have any ideas?" That's my idea. I would encourage you—don't settle for less.

**MULLEN:** Thank you, Gretchen. We're opening this subject for discussion.

**JOHN KULLBERG:** Gretchen, bravo, bravo. Everything you said about Los Angeles I agree with. I do not agree, however, with some of your comments about New York. I happened to make a trip to the Los Angeles Animal Control Department about a month ago. I spent half a day with the general manager, Bob Rush. I invited Bob to come to New York at ASPCA expense, talk with us and give us some ideas, because I've known him for many years, and I like what I see.

The shelter that you talked about in Los Angeles, Gretchen, was built with City rather than humane society funds; $5 million 1987 dollars is being spent on that shelter. City money. No humane society is giving a nickel. The *City* is spending $5 million for a shelter and spay/neuter clinic, and there will be four more built. Some $30 million will be spent by the City of Los Angeles just on buildings. In addition, the City entirely and fully funds animal control programs.

In New York, the City contributes nothing to animal shelter construction and never has. I think that you were exaggerating a bit in reference to New York, and I want to just briefly correct you on that. First of all, in New York, a humane society does animal control to which the City contributes *some* money: the current City contract with the ASPCA is approximately $2 million a year. In addition, the ASPCA nets from dog license

Dr. John Kullberg

revenues about a $1 million. The rest of the money the ASPCA has to contribute—close to $2 million a year. We lose a lot of money in animal control. We can't do it the way we want to, because there isn't more money to do it. We now provide 24-hour ambulance rescue service. The rescue ambulances in Los Angeles took in about 15,000 animals last year. We took in about 30,000 in New York City—injured, stray, helpless animals. Our shelters alone, just the shelters, not the spay/neuter clinic operation, cared for some 80,000 animals, and overall, including animals treated by our hospitals, we directly cared for 127,000 New York City animals in 1986. Our spay/neuter clinics now are doing approximately 6,000 spay/neuter operations annually. Not one farthing comes from the City of New York or from taxpayer dollars to assist us in spay-neuter outreach to the public.

The biggest frustration I have in this city is in dealing with those who say, in essence, "ASPCA, we love to hate you. We love to hate you, because we know if the City took over animal control, the result would be so much worse than the job you do." Now, if only we had the financial commitment of Los Angeles in New York City, or the commitment of Chicago, Illinois, where the city just spent $7.5 million on one animal shelter, or Washington, D.C., where the city built the animal shelter; or other great cities in this country, where the city, in fact, has come to terms with animal control. The city should be funding these programs. How far ahead we would be from where we are now, were New York City to do what city governments do in many other major cities?

When I first came to the ASPCA, I talked a lot about licensing your dog. I ran a *humane society*, but I was out promoting *taxes!* Licensing is important, but it's not what I should be predominantly spending my time on as president of the ASPCA. The City should be doing that. Lately, what am I promoting? I'm out touting the need to put a tax on pet food so we can raise from the $7 billion-a-year pet food industry some $300 million nationally to help the cities find more animal control funds when they tell us: "What do you mean? We've got orphans, we've got widows who need more money!" The animals are at the bottom of the funding list.

**WYLER:** Make the people ask, John.[6]

### Animal Control: How Is It Best Handled?

**KULLBERG:** What I believe you are alluding to, Gretchen, is a referendum. The referendum idea is terrific. If the people ask, Koch and all the politicians will listen. So I think we should come to terms with that referendum idea.

In fairness, New York City, in those boroughs that don't have a shelter, the City is putting in motion an arrangement whereby every borough will soon have some kind of facility.[6] But they think they can pull the wool over your eyes, to be blunt about it. The City says: "We're going to put a shelter in every borough." But what are they actually putting in the boroughs? Did you see what I called the facility in Queens, when they finally funded a facility last year for a couple of hundred thousand dollars? I would not and will never call anything we operate that's not a shelter a shelter, so we called it the "Queens Animal Receiving Facility". It is open eight hours a day, six days a week. It is a depository where you bring an animal. You cannot get an animal there, you cannot spay an animal there, but you can bring an animal for temporary housing before being brought by us that same day to a genuine shelter. *A receiving facility is not a shelter.* But some borough presidents right now are trying to sell that notion to get *you* off their backs. In New York City, where the animal control need is so great, we need a *full service* shelter in each borough. But we've got to find the money for it. I've tried to help the politicians find it through a tax on pet food. Somehow, this money must be found. A referendum might bring greater impetus for them to find it. But until it is, animal receiving facilities represent at least some progress.

One other point: Unlike Los Angeles, ASPCA shelters in New York are never closed. They're open 24 hours every day of the week, and that availability to the public would be an improvement, I think, in Los Angeles. That said, Los Angeles does offer a model program. My wife comes from Los Angeles. She tells me, when I go home and complain: "John, Los Angeles is Los Angeles, and New York is New York. I've seen both, and you just don't have the government commitment to animals in New York City. The public attitudes aren't the same either." We've got to get our act together. Don't let the wool be pulled over your eyes. Understand that the politicians will try to placate, because they've got a lot of demands on them for AIDS research and everything else. When they tell you they're going to erect a "shelter," ask them what they mean by "shelter." And there is no shelter worth the name today that doesn't have a spay/neuter clinic. I will have more to say on this tomorrow.

**BARBARA MEYERS:** I'm the founder and president of the Critter Car Animal Transportation Service, and I have a private practice as a grief counselor for animal-related catastrophic illness and/or death, and I'm the person you're looking for. I need a little help. John Kullberg knows me, and I'm volunteering publicly here to make this happen. I will leave my cards available to everyone. Let's do it.

**ANN FREE:** I have a question that I should ask Margaret Geraghty alone, later on, but I think this is the kind of answer that should get on the record: In the rural areas in Virginia, Maryland, particularly in Virginia, people do not let their dogs run at large, because they are afraid of the game warden or whoever, but it's this constant and perpetual chained, tied out, 'round and 'round, until you've got a trough around. The dogs are fed. Sometimes they get a pat on the head. Sometimes they're left alone *all day*. But what can we do about that? The places are fenced, they are obeying the law, they're not running at large. So, what is the answer? I mean, it's a psychological mistreatment of the animal, of course. But they do have dog houses, and I imagine some of them are sound in the winter, I don't know. I spent some time in the country, and I'm going to be spending more, and I'm going to raise a good bit of hell about this, but I'd like to know how to go about it. Any ideas from anybody here?

**GERAGHTY:** I'll tell you one thing that we have done in Chemung County. We had a local artist draw a picture for us that we highlight in our newsletters. It's a picture of a backyard with a lot of spare tires, junk, a chain around a tree, and at the end of a chain is a flower pot. And the quotation we have is, "If all you want to do is feed and water, get a flower." That helps. That's education. Guard dogs, okay. You better not have a guard dog loose if you don't have any fences.

**FREE:** True, but it's complex to carry it further. I think most of these dogs that are tied up are all there for protective purposes. On an individual basis, I was at one place the other day where the people are not living there any more. The son comes to milk the cows, and the dog is tied all day. And when I drove up there, I was scared. I didn't go back there. But on the other hand, when he came, and I talked to him, the dog was just full of love. But my point is that these are guard dogs in most cases, but not guard dogs in the canine vicious sense that we usually think of. They're guards from the standpoint of alerting people, and people are afraid of big dogs. And not many of them are there just because they like them.

**GERAGHTY:** Education, I still feel, is most important in getting that out to the community.

**FREE:** But what are they going to do with them?

**GERAGHTY:** I suggest individually, if the individual is that concerned about a particular dog, contact the nearest SPCA, the nearest person who is licensed to take care of possible cruelty to animals.

**FREE:** I don't think there's any cruelty, except psychologically.

**GERAGHTY:** That *can* be.

**FREE:** Well, it's not in the law.

**GERAGHTY:** That *can* be, if it's neglect.

**FREE:** And you're not going to get the country folks to buy that one.

**GERAGHTY:** You're right.

**FREE:** Horrible problem, and almost insoluble, so I think we ought to get our minds going on this, and that's why I'm inviting anybody who could come up with something.

**MULLEN:** Thank you. I'd be very interested, since we do have a representative from the mayor's office here—Mr. Martin Kurtz—in hearing his comments on the suggestion made by Gretchen Wyler about the possibility of seeking a referendum related to City animal control. Would you be willing to comment, Mr. Kurtz?

**MARTIN KURTZ:** As you know, or should be aware of, the ASPCA in the City of New York recently helped get a State bill passed in Albany that would allow the mayor to elect to use an alternate provider for animal control services presently provided by John Kullberg's society, and also gives John Kullberg the option of letting the City know, in a certain period of time, that they no longer want to provide the City's municipal animal control services. However, John and I, although we're friends, we have a lot of disagreements as far as what the "A" is and what the "A" is all about, and what the City is, and what the City is all about. Being in the Department of Health and going back historically to when there were farms and when there were problems with the pasteurization of milk, the City always had veterinarians and some animal-related services, also because of the rabies problem in New York City at that time.

The ASPCA and the City contracted in 1977 and formed what I felt, and I still feel, is a partnership of one of the best animal control services in New York City, and I also feel, and I'm not criticizing any other program, but New York City has the expertise and the experience with John and the people from Bide-A-Wee, with other New York City people, to make any kind of the best animal control service possible in New York City. I'm not criticizing Bob Rush or any other municipal program, but I'm damned proud of what we have here in New York City. We are going to seek to see if there are other providers to take care of certain municipal animal control services, basically, those would be the shelter care of animals and the animal rescue and pick-up of stray unwanted animals. The other services that were mentioned by the panelists as far as animal control, rabies surveillance, animal bite reporting, vicious dog ordinances, those are all really part of the Public Health of the Department of Health of any municipality, and it wasn't clear: Do you both represent departments of health or police departments, or what is the city agency for which you work?

## Animal Control: How is it Best Handled?

**GERAGHTY:** We're a private, nonprofit organization, but we do work very closely with the Public Health Department in the City of Elmira.

**REMICK:** In the city of Ithaca, in Tompkins County, we work very closely with the Health Department, but we're not directly a part of that agency. We quarantine animals that have bitten somebody.

**KURTZ:** In New York City, I think the relationship between the Department of Health and the ASPCA is right now the appropriate relationship. If the City can seek alternate providers who would be qualified, capable vendors to provide animal control, we would certainly listen to them, and we are going to advertise for that in the next month or so. So the direction of animal control in New York City really is at a crossroads with the ASPCA wishes, with the proper City funding to continue ASPCA being the City's partner in animal control, or that other providers exist that can provide a quality service that New York City would be proud of.

**MULLEN:** By other providers, Mr. Kurtz, do you mean other humane organizations?

**KURTZ:** The law specifically states that any provider must be a Not for Profit corporation under New York State Not for Profit corporation law, and must also be a humane society, not involved in, I forgot the exact wording, of the sale of animals for research, for education, for training. It would be basic humane society qualifications.

**MULLEN:** I see. But as to Gretchen's specific suggestion about a referendum, what are your thoughts on that?

**KURTZ:** Everybody's free to do whatever you want to do, you know. We'll listen. The City does listen. Thank you.

**MULLEN:** Thank you, Mr. Kurtz.

# WHO IS RESPONSIBLE FOR PET OVERPOPULATION?

**MULLEN:** I'd now like to introduce our next speaker whose topic is, "Who is responsible for pet overpopulation?" A change in the program involves the absence of someone I know that you've all looked forward to seeing, Phyllis Wright, who won't be able to join us today. Her time will be used by Ann Cottrell Free whom I will introduce later. She will be telling us about a tragic aspect of animal overpopulation in Vieques, Puerto Rico.

Barbara Cassidy is a colleague of Phyllis Wright's, and works out of the same office at The Humane Society of the United States, where Barbara is Director of Animal Control and Sheltering. Prior to taking that position about three years ago, Barbara was my predecessor in the position of the Public Affairs Administrator of New York State Humane Association. She was the first full-time paid employee for the New York State Humane Association, a position that she held with distinction for five years before going to The HSUS. We've continued to work together closely. Before taking a position with the New York State Humane Association, Barbara was the manager of the Ulster County SPCA shelter. She had to work under very adverse conditions there because of an inadequate physical facility. Barbara showed what can be accomplished in spite of an unsuitable building, and she helped the Ulster County SPCA gain a fine reputation as having a well-run animal shelter with excellent policies.

**BARBARA CASSIDY:** What I'm here to talk about is who is responsible for pet overpopulation. I think that it is a given that we do have the problem. I want to say something that's a little contradictory to what some of your other speakers have said, and that is, I don't think the problem's increasing. It *is* decreasing. The numbers of animals in shelters and the numbers of animals that are subsequently killed are going down. There are no real hard, fast statistics, because there are no national reporting requirements, but one thing that the HSUS does is conduct surveys. We are in touch with animal shelters across the country on a regular basis, and every few years, we do a fairly comprehensive survey. Ten years ago, the numbers of animals that were going through shelters were estimated between 15 and 18 million and we were probably euthanizing 14 million. The current estimates today are that shelters are receiving about 13 million, and we are euthanizing about 7.5 million. That is still a national tragedy. That is, in our view, a disgrace.

So who's responsible? You'll probably all say, "The irresponsible pet owner." We've made some progress, but how have we done it? Chiefly, it is through what we call the "L.E.S." program. What it means is: "Legislation, Education and Sterilization," and your speakers this morning all alluded to one aspect or another of what that means, but I want to elaborate on it a little bit more.

Communities across the country that have implemented responsible animal control ordinances have seen numbers go down, and when I say, "responsible," I include the enforcement of those ordinances. You have to have a good ordinance before you can do anything about the overpopulation in your community. And what is a responsible ordinance? It basically prohibits free-roaming animals, includes differential licensing, and by that, I mean there is a incentive for sterilization, that you get a very inexpensive license for an animal that's sterilized and a very high price for a license for an animal that is unsterilized. It includes, of course, the mandatory sterilization of any animal that's adopted from a shelter, and rigorous enforcement of all aspects of the contract. In this day and age, if any shelter cannot confirm—and I don't mean collect a deposit that you sit on—the sterilization of more than 75 percent of the animals it releases, it's unconscionable; you're not doing your job.

There are a lot of cities that have implemented good ordinances and have implemented the education and the sterilization components. Gretchen talked this morning about Los Angeles, which has been one of the cities that we hold up to the rest of the country as having an exemplary animal control ordinance and enforcement. What Gretchen didn't tell you is that, ten years ago, or a little over ten years ago now, they opened their first spay/neuter clinic. They were handling in excess of 150,000 animals at that time. The real big difference with all of the marvelous programming that Gretchen talked about is that the number of animals handled has been cut in half. It's about 80,000 now. That is still a tragedy, but it is a major difference. It is what every city should be aiming for, whether it is rural, metropolitan or suburban. Other cities that have had similar successes include: Santa Barbara, California, a small city which, in 1972, was handling almost 9,000. In 1984, they're handling 3,000. Humane Society of the Huron Valley in Michigan, was handling 16,000 ten years ago. Now they're handling about 8,500.

Education: It is imperative that every humane society and animal control facility participate in humane education in the community. This includes education on the importance of sterilization. You *cannot* make a difference unless you are changing the attitudes and values of those in your community.

Sterilization: It is important to make the sterilization available at low cost, and educate the public as to the benefits, and then provide the incentive with differential licensing. This approach works. We can look at these cities across the country, and we *know* that it works. Some states have also enacted legislation along similar lines. California, and most recently, New Jersey, have spay/neuter public assistance, Florida has mandatory sterilization and low cost, as do Illinois and Oklahoma. When I say state legislation, I'm talking about providing low-cost sterilization *and* making it mandatory that animals that leave shelters, whether they are private, nonprofit or municipal, must be spayed or neutered.[1] It's not just collecting a small deposit. I submit to you that if you are collecting $10 or $15, the potential pet owner is going to say: "Hey, I bought a cheap dog," and the sterilization is most likely not going to be done.

What is absolutely necessary is the follow-up. First of all, $10 or $15 is not enough. You should collect at least one-half the cost, and you must follow up. You cannot simply take that money and then, as Gretchen mentioned, maybe put it into general funds. Now, they found a creative way of using that money to serve the same purpose in LA, but what you often see, or we see, on a national perspective, is enormous sums of money accumulating or being put into the general fund, that are not being used for the purpose they were intended for. It's imperative, particularly on a shelter-by-shelter basis, that you not only collect that money, but you see that the sterilization is done. There are a number of ways to do that, and I can answer questions about that when the time comes.

We can see what works, so why do we still have an overwhelming problem? Even though we've cut it in half, we're killing 7.5 million animals. Why do we have 13.5 million coming through shelters every year? The HSUS estimates—and this is probably a more staggering figure—that 70,000 puppies and kittens are born each day. It's a little bit closer to home when you look at those figures. And only 20 percent of those animals will ever know a warm and loving home. The rest will either be euthanized or suffer horrible deaths by the side of the road, from disease, be hit by cars, or endure other abuse. A dent has been made, but we have not solved the problem. One of the things I'd like to do before going a little further is ask you who's responsible. Has anyone got a thought or a suggestion on that? Who's responsible for the pet overpopulation?[4]

**VOICE:** Part of it's the American Kennel Club.

**CASSIDY:** Okay, why do you say that?

**VOICE:** Because they allow the registration of puppies from puppy mills.

**CASSIDY:** Good. Actually, we're talking about two things. We're talking about the AKC *and* puppy mills. I have some very encouraging news about the AKC. They've recently installed a new director who has made a public statement that they are going to try to "clean up their act." I hope so, because they are certainly part of the problem. When you buy animals in a pet shop that are AKC registered, that pet shop animal has most likely come from a puppy mill. They *are* contributors. You're not buying, necessarily, a quality animal. What a responsible breeder will do, and I mean a *very* responsible breeder, is to breed very, very rarely and only to enhance breed traits.

Something else that's interesting that you might all want to know about is a study done by the Humane Society of Santa Clara, in California. They did a study for two years in a row on the numbers of purebred animals in their shelters. They found that 22 percent of the animals they received in that shelter were purebred animals. That's a lot. Forty-six breeds were represented. Sixty-two percent of those were free-roaming, unneutered males. We've always said we know that purebreds are pretty well a dime a dozen in animal shelter, and this agency has proved it.

The concept of merchandising of animals has been brought up. I have to agree that appeals such as: "Empty out the shelter at Christmastime; give these animals a home" don't work. We've had people talk this morning about what quality adoption means, including pre-screening potential adopters. Those that are unsuitable are refused. Quality adoption includes the requirement that new owners sign a binding adoption contract that includes the sterilization, and that you follow up on that sterilization to make sure that it's done; that you not just *say* that you will take the animal back, that you *insist* they bring the animal back if it is no longer able to be a part of that home. We agree that the merchandising of animals, the promotion of adoption, is a serious part of the problem, because it puts animals out in the community who are going to reproduce and are most likely not going to be responsibly cared for. They are essentially the animals that are recycled right back into the shelter again, because they've been free-roaming or they aren't properly cared for, and are maybe even victims of cruelty or abuse. When we talk about the quality of the adoption, and what that means in the long term, it's the quality of life for that animal for the rest of its life. That's something that's really, really very important.

**ANCHEL:** I was wondering about the binding contract. What does it mean legally?

**CASSIDY:** A legally binding adoption contract? We've got two things that we can talk about here. First of all, the ordinance: if you have mandatory sterilization in the ordinance, your contract has a lot more weight in court because they have violated the ordinance. You can reclaim that animal or you can see to it that the animal is sterilized. If it is not part of your ordinance, it can still be successfully taken into court as a violation of a legally binding contract. The contract is not complete until all provisions have been met, and the language of the contract is obviously very important. You can't just accept a deposit and say that the animal must be spayed. A specific date is necessary. If you're adopting puppies and kittens, you must specify the appropriate age.

To make the adoption contract enforceable, HSUS recommends the incorporation of a "Liquidated Damages" clause. Basically, the liquidated damages clause sets an amount that the adopter will have to pay if the shelter brings a successful suit against him or her for failing to comply with specific provisions of the contract. To win the case, the shelter need only prove that the adopter breached the contract. The clause includes a provision for the adopter to pay attorney's fees and court costs for enforcement.

**MULLEN:** I think the main problem with contracts is that they're often considered unenforcable. In fact, they may be enforceable at sufficient cost, but few organizations are willing to become involved in court and to deal with the frequent indifference of judges.

**CASSIDY:** You're absolutely right. Sometimes it's too costly to take a case like that into court. What I suggest before you even get to that point is that you need to properly pre-screen the potential adopters and advise them of what is necessary; that you have collected a sufficient deposit that they will not forego it; that you have a confirmation system in place so that you have only a small number that you are then dealing with that may not comply with the contract. But there are other steps to take. I shouldn't have been so quick to say that you can take it into court. There are other steps to take, such as follow-up notices. An attorney can help you design a letter that is sufficiently threatening that it will get another 20 percent of those unconfirmed sterilizations done. You can narrow those down to a very, very small number. I wouldn't say to just quickly haul them into court for a violation, but rather follow a number of steps to ensure the surgery is done.

**PAMELA MARSEN:** I think one of the problems is shelters that allow volunteers to be the ones who screen adoptions and conduct follow-up. Volunteers may or may not be accountable, may or may not be responsible. Many of them are wonderful. But people come and go. They have nothing to hold them to in the four months between when the cute little puppy or kitten is adopted and four months later, when they become sexually active. A lot of volunteers aren't there any more, and we believe that follow-up is extremely difficult and time-consuming. We think shelters have just got to assume that job is at *least* as important as paying somebody to log in the animals as they come in.

The other thing I think is very important is neutering the animals before they go out. And I don't care what the objection is, you know, that it's going to cut into the adoption rate. Too bad! [4]

**CASSIDY:** Good point. And there's no reason, if an agency has the wherewithal, not to enact such a policy. Many shelters do. If you can sterilize the adult animals before they leave the facility, that's marvelous, but not every shelter can do that, and you can set up very workable programs that rely on the pet owner to have the sterilization done prior, or at the appropriate date. But again, it requires pre-screening and adoption standards, policies, procedures, and that they be carried out to the letter.

I don't *entirely* agree with your thoughts about a volunteer program, because I have seen some excellent volunteer programs. There is a big difference between allowing someone to come in and sort of fool around with the dogs and cats and handle your paperwork or, as Gretchen talked about earlier, do your lost and found. On that basis, no, it doesn't work. But I've also seen volunteer programs where the volunteers are rigorously and properly trained, where they commit to a certain period of time, an actual docent program in place that includes training and commitment. When you set something up that well, when the volunteers have a great sense of commitment and accomplishment, and you have some continuity, they can be a very great asset to the agency. But there are all those provisos, so it *can* work, but not always. Any other thoughts on who's responsible?

**VOICE:** Casual breeders.

**CASSIDY:** Casual breeders, right.

**JULIE MOSCOVE:** I think you've all said good things here, but I think that you're missing the point in a lot of different aspects. The person that is responsible for what's going on today with our pet animals is the irresponsible pet owner. Any way you look at it, there's only one way to cure the problem, and that's to get to these irresponsible pet owners and make them responsible. Now, how you can make them responsible is by tattooing these animals before you put them out for adoption. In all the cities, there are abandonment laws that can be enforced, but unless these animals are linked to these people that you're giving them out to, you have no control whatsoever.[5]

Owners can be prosecuted for abandonment. This doesn't have to be done by the shelter. It can be turned right over to the city, and there will be no effective controls throughout this country until a program like this is started. I happen to have a program which is available to shelters free: give them the equipment, put out your animals on an adoption basis tattooed. The owners are now responsible. Now if anybody has checked with the British Columbia SPCA in Vancouver, it's a program that works. In France, mandatory tattooing—it works. In Alaska, mandatory tattooing—it works. When are we going to wake up? I mean, the people that are irresponsible, give them the free spay and neuter, they're not going to do it. Who's going to tell them to take their animals in? Those people that are conscientious, yes, they will do it. And those that want to have it done will pay less to have the licensing rather than pay more and not have it done. But the irresponsible pet owner is not going to do anything. He's just going to open his door, he's going to let his dog out, or he's going to have his dog out, because it never even gets in. And this is the story. Until you get to these people, forget everything else. Get to the people.

**CASSIDY:** Thank you. I don't think I want to forget everything else, though.

**MOSCOVE:** Everything goes with it, but that's the most important thing, really, making them responsible, because they're the ones that are guilty. It's got to be a right to own a dog, really.

## Who is Responsible for Pet Overpopulation?

**CASSIDY:** That's an interesting concept, too.

**VOICE:** I think that a lot of the problems or the solutions that you have suggested are really treating the effect, not the cause, and I think that you've really got to get down to education.

**CASSIDY:** I was waiting for somebody to say that.

**VOICE:** The problem is really the schools, I think, because the generation that's now parents is hopeless. I'm a veterinarian in private practice, and I'm telling you the things that you hear, it goes right over their head, because it's too late for them. They say, "This is our 27th kitten, and we've always found good homes for them." You try to explain to them that that reasoning is fallacious. They don't hear it. The place to start is in the schools, by requiring an education of these children, so the children say, "Mommy, our cat should be spayed." or "Our dog should be neutered." That's the only way. You can't reach them. I try on an individual basis, and perhaps in a lot more intensive way than shelters can, and I can't get through most of the time.

**CASSIDY:** Thank you, good point. I appreciate that, and I think you're absolutely right. I said: "legislation, education and sterilization," and I was waiting for somebody out there to mention education. It's absolutely, critically important to work on changing attitudes and values. I want to expand on it a little bit and say that it is not just the kids that we have to educate, although that is where we should spend a great deal of our time and effort. The humane education division of HSUS—the National Association for the Advancement of Humane Education[6a]—spends a lot of time looking for ways to creatively educate children about responsible pet ownership and spay/neutering. A new program that's being launched, which is very much what you're talking about, is called the Humane Chain. It's getting the kids to do those little paper chains that they love to play with in grade school anyway, and, as a group, building long chains and putting them in their communities' veterinary hospitals and public places, along with the accompanying public education. The point is that the kids are talking it up and anyone who takes a link off this chain is committing to have an animal sterilized. And they are, at the same time, obviously getting the message on what spay/neuter is, what overpopulation is. The children are taking that message home to their parents, to their family members.

Education is critical, but it cannot be aimed just at the children. There are a number of local agencies that are doing education campaigns, but I submit to you that humane societies are not doing enough. We talked earlier this morning about animal control being the responsibility of the municipality. I firmly and strongly believe that. That is *their* job, and they can do it properly and they can do it humanely. It should be adequately funded in the budget. Have an advisory committee as the liaison between the humane community and the animal control community. Fostering that kind of program is what the humane society should be doing, and then spending their time on educating the public.

There are some marvelous education programs out there. I brought a few just to show you.

Massachusetts SPCA is launching their local spay/neuter campaign, and they've got some of the cutest stuff I've seen in a long time. They're going to have posters in Massachusetts newspapers all over the state. One ad says, "Sam enjoys walks in the park, playing ball, nights out with the boys and casual sex. And he's only three. He's also one of the most sexually active house pets responsible for the 13.5 million animals that go through shelters in this country."

Another one: "There's more than one litter problem in your neighborhood." Here's another one: "Surprise! Your four-year-old is the grandfather of 173."

These are great, and there're going to be in newspapers all over the state of Massachusetts. This is going to be educating the public, because they don't know what's going on. We've often hidden the euthanasia statistics in animal shelters, which I am firmly opposed to. I think you've got to tell the public, put the blame squarely where it belongs, on the irresponsible owner. However, it is also up to us to educate them. They're not going to change their attitudes and their values without the education and without us pushing, and without us enacting ordinances that make a difference.

Another really good campaign, in Washington, D.C., has been

## Who Is Responsible for Pet Overpopulation?

launched by the Washington Humane Society, which by contract operates the animal control program and spay/neuter clinic in the District of Columbia. They've gotten Washington Redskins player Mark May to do a poster with his three shelter animals, all of which are spayed and neutered. Posters are placed throughout D.C. that say, "Everyone wins when your pet is spayed or neutered at the low-cost spay/neuter clinic." They want to increase the numbers of animals that are done in that facility. They've done 3,500 surgeries since they've opened up, and it is a relatively new clinic.

The HSUS is launching a national campaign called, "Be a P.A.L., Prevent A Litter" campaign, which we hope also will educate, and I'm not here just to show it off: I want your input as well. I've got two posters that I'm going to show you, and a "Close-Up Report." We have done a lot of work on overpopulation, and a lot of what I do with shelters on a regular basis concerns that, but we are renewing the pet overpopulation campaign. We want to have national impact. We want to emphasize that we want you folks on the local level to take the message and go home and do something about it. One of the things we're going to do is make available one of each of these two posters, free of charge, to every shelter in the country. You will be able to order others, and the reason I want to show them to you now is I want to get a little input from you.

What you're going to get with this poster will be a fact sheet, public service announcements and some material on what you can do in your community to educate the people who are responsible. Feed back to me what else you would like, because we are in the process of putting together a kit, and if there's more that we can do for you to help you on a local level, I want you to tell me about it.

First, this is the "soft" version. And what the campaign is called is, "Be a P.A.L., Prevent A Litter." Most of you, I'm sure, recognize Willard Scott. Willard is launching our campaign.

Now, the other poster is what we call the "hard" copy. This will be included with the "Close-Up" which talks about the current statistics and what the problem is in the U.S. This is a little puppy photographed in front of a rather large pile of already euthanized puppies and kittens. It's very dramatic, but this is what happens in a shelter, as we all know who work in a shelter on a daily basis. It says, "When you let your pet bring unwanted animals into the world, guess who pays?"

This will go out with a "Close-Up Report" with the current statistics, an outline of the problem and some of the solutions.

Other resources are available, such as a model animal control ordinance and an animal control program, and I encourage you to get them from us. We're talking again now about the ordinance and responsible animal control, which in my view, is the job of the municipality. I don't want to restate it too many times, but the humane society has an awful lot to do with preventing cruelty and educating the public. Animal control *should* be and *is*, in many communities, the responsibility of local government. How do you get those ordinances passed? We've got resources and information that will assist you in talking with your local government. New York City may need a referendum. I don't know if you're going to need that in every community, but I strongly urge you to work on the ordinance. I think we're responsible for making those changes. We're not responsible for the problem. I know that every one of us is awfully tired of being blamed for the animals that we euthanize. We *are* responsible for making a change.

**VOICE:** I think the phrase, "We do not destroy" gives the uninformed public the illusion there's really not a problem in placing a dog, and gives people free rein to feel like they can go out and let the animals go ahead and breed.

**CASSIDY:** I agree. I generally say the numbers of animals we kill, and that is the hard fact, that's what we're doing.

**VOICE:** When you see advertisements in the paper, "Bring us your puppies and your kittens," that kind of gives the public the idea that . . .

**CASSIDY:** What do you mean, "Bring us your puppies and kittens"?

**VOICE:** North Shore.

## Who is Responsible for Pet Overpopulation?

**CASSIDY:** Oh, when they're soliciting. Okay, I understand what you're saying.

**DR. HERBERT RACKOW:** I think you said a shelter that neuters 75 percent of its adoptions is acceptable?

**CASSIDY:** Yes. That's getting close.

**RACKOW:** But even that is not really good enough. You pick up one hundred strays. You neuter 99. You adopt out one fertile animal. That one fertile stray over its lifetime might have 100 offspring.[4]

**CASSIDY:** I agree with what you're saying, but I think there's also some other reality that some animals or some people will slip through the cracks. I'm saying if you do any less, that you are not on the right track. If we could get 100 percent confirmed sterilization, that *is* the ultimate, I agree with you a hundred percent on that.

**PAMELA MARSEN:** I'd like to speak on a subject that's very close to my heart. I've wanted to remind you that at the HSUS convention that was held in Chicago two years ago, the incoming president of the American Veterinary Medical Association and his cohorts announced that they didn't think it was necessary to stop animals from breeding that, as a matter of fact, we could stand a mild increase in the numbers! So don't forget who your friends and opponents are.

**CASSIDY:** On the other hand, the AVMA has a policy statement which may be just so many words, but they are on record, in writing, in strong support of cooperative spay/neuter programs.

**MARSEN:** Okay. Also, in terms of your print campaign and method, it's very nice, but I think we have to be smart. I read somewhere that I think it was ten percent of the U.S. population subscribes to and/or reads newspapers. Also, talking about children, the major way children form their attitudes is they do it not in schools, it's not in going to animal shelters and seeing posters, it's in watching TV, and we're not on TV.

**CASSIDY:** I did not tell you about the entire campaign, but we will not just be doing these posters. We have a film that we have just finished that will go on cable TV and will also be available in schools on video. We are going to hit other avenues. We also have appropriate other materials for children.

**MARSEN:** Okay, because I teach humane education, I go to the schools and I always ask the teachers to have the students write me a follow-up note to tell me what they learned. And I'm very sad to say that not many do. But I think it's a pretty good program. Not that many even mention spay/neuter, despite the fact that I spent an hour and a-half showing them films.

**CASSIDY:** I do want to say that you can't do it in a "one shot" visit. Humane education has to be ongoing and repetitive, and it can't be *your* job alone. The *teacher* has got to have the tools and the resources to have humane education in one form or another on a *regular* basis. You're not going to change any attitudes or behavior with a one-shot humane education session.

**MARSEN:** Right, but I really do want to encourage you to push for being on television, because it's so very important.

**JACQUELINE BULLETTE:** It's great what Pam was talking about, about television, but I think the last thing the animals need is another flyer, or another flyer on a shelter. What I would like to see the societies who have plenty of funding sitting around, do, is the following: There are something like 10,000 community papers in the United States, 600 college newspapers alone. I would like to see run in these papers, on a daily basis, the Friends of Animals 800 telephone number for the low cost spay/neuter certificate.

## Who Is Responsible for Pet Overpopulation?

**VOICE:** I have two comments. The first one is that you asked the question: "Who is responsible?" and we've listed a lot of subgroups, and in fact, we even excluded ourselves from being in one of those subdivisions. But my comment is this: I believe *everybody* is responsible, not just pet owners, not just irresponsible pet owners, not just pet owners who don't have their animals spayed or neutered, but I believe it's everybody, even non-pet owners. The reason why I say that is that everyone, everybody in the community, suffers from pet overpopulation. They suffer in terms of tax dollars being spent to try to control uncontrolled dogs. They spend money in trying to pay for insurance premiums because of dog bites, because people don't take care of their pets. So I believe it's everybody's responsibility, and if, let's say, I don't have a pet, and I see my neighbor having a female dog, and every year, it's having litter after litter after litter, and I say to myself: "That isn't my problem. You know, that's someone else's problem. They should really go have that taken care of, but I'm not going to say anything."

So I believe it's everybody's responsibility, and we can say, "Hey, it's more the responsibility of people who don't properly take care of their pets," but nonetheless, I don't think any one of us can close our eyes to the problem.

**CASSIDY:** No.

**VOICE:** Now, the next comment, and that is in regard to education. I think we need to go beyond education. We have to educate first, but that education has to turn into action by people. We've got to get people to spay/neuter. We've got to get people to keep their pets at home. We need to get people to take their animal to a veterinarian. We need to get them to have the idea, that when they have a pet, it's a pet for a lifetime, not for just as long as this pet suits their lifestyle. So we need to go beyond education, and turn that into real action.

**CASSIDY:** Thank you. You've practically stolen my closing remarks, because that is going to be my challenge to each one of you in the audience. It's renewed effort. Go back to your community and start to take an activist approach to the problem. I think what you have to do is work on the education, work on the ordinance, get involved with local government and have them provide a responsible animal control program in the community, that is funded properly so that they can do the job. I think that we need to make next year one of renewed effort, of greater commitment to the solution of the overpopulation problem. Let's stop taking the blame for killing the numbers of animals that we do, and start reducing it.

**ANN FREE:** An idea: Why not act on what we're talking about right now, getting it into, for example, the *National Enquirer*, maybe *The New York Times*? Before this meeting is over, have a poll, a vote: "Who's responsible? The New York State Humane Association had a ballot." And we'll poll everybody sometime tomorrow and have maybe five different subheads. It might be church, state, schools, and so on. We can think up the format, but I would say get some publicity out of the whole pet population explosion. Everybody likes polls, so I suggest that, and I'd be happy to work out any details with you.

**CASSIDY:** Thank you.

**VOICE:** I'm not an animal control or humane professional. I'm a laborer, and I'm here because I love animals. I've learned a lot. When you showed that poster before, of the puppy in front of the pile of euthanized animals, I could hear a wave of shock go through the room. I think that's a good poster. People in the public, like me—you have to deal with this problem every working day—people like me can drive by the shelters where this is all going on and never go into a place in our lives. The members of the public need to see that. People may say it's shocking. There are some fairly shocking pictures of AIDS patients.

**CASSIDY:** You're going to see that ad eventually in *U.S.A. Today*.

**VOICE:** Good. And, you know, all that is done in the name of education I'd be glad to see it.

**CASSIDY:** Thank you very much.

# Control of Feral Cats

**MULLEN:** Our next topic is about an activity with which I've had a great deal of personal experience. But I've found it's a subject that's been virtually overlooked in most animal shelters. The shelter in Kingston, New York, with which I'm affiliated—the Ulster County SPCA—operates, primarily through volunteer services, a program for the control of feral cats. It involves the live trapping of cats.

AnnaBell Washburn is our next speaker. She is very active in New York City and on Martha's Vineyard in humane education and in another important endeavor, the trapping of feral cats. AnnaBell has gained her expertise both on a practical level and through consultations with the Universities Federation for Animal Welfare in Great Britain. She has implemented the methods she's learned for the control of feral cats on the islands of Virgin Gorda and Martha's Vineyard, as well as in various sites in New York State.

**ANNABELL WASHBURN:** Thank you, Samantha. I'd like to know how many of you have ever been involved in trapping colonies of cats, stray cats. Would you raise your hands? Oh, well, in this friendly gathering, I'm not alone. I think, to begin, I'd like to show you a video on the control of feral cats. It's the UFAW video.[6b]

**ANNOUNCER [videotape]:**

> *Feral cats can be successfully controlled without necessarily killing the cats. This presentation describes the recommended methods of trapping, neutering and returning feral cats to supervised sites, methods that have been developed by the Uni-*

versities Federation for Animal Welfare at the request of the Cat Action Trust. Further details are given in the UFAW publication.

For many years, small groups of feral cats have been welcome living in and around farms, where they provide a free rodent control service. In recent times, stray or abandoned cats and their offspring have gathered together in urban areas. Wherever there's a source of free food, they soon start to breed and to form family groups. These are now found in such places as the public parks, hospital grounds, dockyards, and industrial or housing estates, usually somewhere where a kindly person starts putting out food for them. However, in some places, they are not so welcome, and they can also cause problems. They tend to make a lot of noise, particularly at night. Toms fight over territory, queens call for mates when in season. Both may fight over food supplies. Not only do they fight, they can also smell. Tomcats mark their territory by spraying pungent-smelling urine, a warning to rival tomcats, and also unpleasant to humans. Smell is often the first complaint received from the public by the local authorities. Cats also carry fleas. These are a problem mainly to the cats, but fleas jump on people, and they cause irritating bites. In hospitals, fleas may be carried into the ducting systems. They have even been found in operating theaters. Although in some places, small feral cat colonies may be tolerated or even welcomed, when numbers increase, they may become an unacceptable nuisance. How, then, might we best solve this problem?

There are really three options: leave well alone, try and eradicate the whole group, or establish a properly managed colony.

First, if colonies are left alone, their size will be regulated by natural processes. When there are too many cats for the available food supply, some fit adults may move elsewhere, and weaker ones and kittens too young to fend for themselves will surely die of starvation. The chief method of regulation of the population will be the death of kittens. Most kittens will die from diseases such as cat flu and

## Control of Feral Cats

*enteritis. Perhaps only one or two from each litter will survive. The spectacle of sick or dying cats is most distressing to all concerned.*

*More humane methods are obviously required. For many years, feral cats have been dealt with like any other pest problem. When they become a nuisance, attempts have been made to remove the whole group. Cats have been shot or poisoned. But shooting is dangerous, and it is illegal to poison cats in England and Wales. The more usual method is trapping and then killing by an injection or in a chloroform chamber. However, it is time-consuming and expensive to achieve eradication of the whole colony. Usually, some cats escape and other cats soon move in, and breeding starts once again to fill the vacant territory, so this method doesn't solve the problem. Furthermore, such a process is usually strongly resisted by local animal-lovers.*

*This leaves the option of management of a supervised colony. This method needs a carefully prepared plan involving the land owner or administrators, the feeders, local residents, and veterinary surgeons. Cats, then, have to be trapped. Those that are kept have to be neutered, then either homed or returned to the supervised site, and those who are too sick or too old, and not to be kept, put down.*

*Traps may be hand–operated. This is done by the trapper hiding nearby and watching until the cat enters, before pulling the door shut. This is selective. The trapper can decide whether or not to trap the particular cat which has entered, but it requires constant close vigilance, and the trapper's presence may deter the cat. Automatic traps are therefore usually used for large-scale schemes. Cat food, tasty bits of chicken or sardines, are placed at the entrance to entice the cat inside to a larger amount at the far end. The cat steps on a treadle which releases the trap door and this is then held shut by a restraining bar. Trappers can position themselves some distance away to allay the cats' suspicions, but cages should not be left unchecked for more than two hours.*

*Once a cat has been trapped, it should be trans-*

*ferred to a squeezeback carrying cage. If the cat is alarmed, throw a blanket over the trap. The darkness will help to calm it down. Never attempt to handle a feral cat. It will defend itself with claws and teeth, and can inflict serious damage. To transfer the cat, place the carrying cage tight up against the trap door. Then raise both doors and encourage it to move across by blowing or gently prodding with a twig, or by leaving a jacket over the trap. The cat will move from dark to light. The cages should then be taken to the veterinary surgery, remembering, of course, that as part of your plan, you have made arrangements for the veterinary surgeon to receive them.*

*Depending on the time of year, there will usually be kittens in every colony. A special effort should be made to find them. This is especially important if you've trapped a lactating mother. Catching kittens is not always easy. They hide in all sorts of awkward places: pipes, boxes, old cars, up trees. The older the kittens, the more difficult they will be to find and catch. Kittens up to about eight weeks of age can usually be easily tamed and will make excellent pets after veterinary treatment. Some may be too sick or injured, and the vet may decide it's kinder to kill them. Otherwise, after treatment and the surgery, it's worth trying to find them good homes. This kitten is given treatment at the Mombasa Clinic of the Kenya Society for the Protection of Animals. If you do manage to find homes for the kittens, remember to tell the new owners to have the cat neutered and vaccinated when it's old enough.*

*The veterinary surgeon will first wish to give adult cats a thorough examination. The squeezeback cage has been specially designed with a movable panel to one side. With this, the veterinary surgeon is able to anesthetize a cat with an injection through the bars of the cage without actually handling the animal. Once unconscious, it can be examined and assessed for surgery. It may be better if sick, injured or very old cats are humanely destroyed straightaway. Those that are fit are then neutered. The veterinary surgeon should be asked to spay female cats through a flank incision. Midline incisions are more likely to drag over the ground or become infected when the cat is released. He should also be asked to remove the ovaries and the womb when spaying, and to abort any fetuses from females which are pregnant. The surgeon should be asked to sew up the incision wound with absorbable sutures which do not have to be removed.*

*Male cats are neutered by castration. Neutered cats should be permanently marked to make sure they're not presented for surgery a second time, by mistake. The veterinary surgeon can do this by removing a one-centimeter tip from the left ear while the cat is still anesthetized. This is painless and harmless, so long as the cut edge is sealed. All cats should be given an injection of a long-acting antibiotic to counter the risk of infection after surgery. In countries where rabies is endemic, cats should be vaccinated against this disease. It's not usually necessary to vaccinate feral cats against cat flu or enteritis, because having survived to adulthood, they will usually already have acquired immunity. Cats may also be treated against tapeworms at this stage. Each cat should be sprayed or dusted with anti–flea preparations, and any minor wound or ailment can also be treated before it's placed back in its cage. Cats should be placed in a warm, quiet place to recover. And after 24 hours, they should be fully awake and alert.*

*Once the surgeon has declared the cat fit, it should be taken and released as close to the point of capture as possible. Most cats will not wait to thank you, and will rush away as fast as possible. Before leaving the site, be sure you have left food and water. Boxes may also be provided. Newspapers and old blankets make ideal bedding, and this should be replaced every two weeks, and once a week, sprayed or dusted with anti–flea preparation. This will keep them free of parasites.*

*Of course, this costs money. The initial cost may be quite high, but in the long term, it is cheaper than repeated attempts to eradicate an ever re–forming colony. The cost may be broken down into some main headings: equipment—the cost of traps, cages and bait; transport—petrol for the use of car, if needs be; veterinary services—the surgery, drugs and treatment of cats and kittens; post–operative care—food and keeping in a warm shelter; on–site care—feeding box, cat food, sleeping box, bedding, anti–flea treatment, and worming tablets. The total obviously depends on the size of the colony. Nevertheless, the benefits far outweigh the costs. The benefits to cats of such a management program are obvious. They continue to live a happy life after neutering, and often improve in condition. Cats become friendlier towards each other and towards their feeders. The main benefit is long–term control. The problems of smell, noise and fleas are alleviated to a large extent, and of course, neutered cats no longer produce kittens. The colony will remain stable in numbers. The cats will defend their territory against intruders. As cats die through old age or accident, others may join the group. These should be trapped and neutered as well. So, with a stable, well–managed colony, the cats will continue to give pleasure to the local animal lovers. And that concludes this presentation.*

**WASHBURN:** Now, I'd like to give you a practical application of the information that we just heard. About three years ago, I attended a World Society for the Protection of Animals conference in Boston. At that conference, Peter Neville, Bachelor of Science and animal behaviorist, gave a presentation pretty much like the one that you have just seen, describing the Universities Federation for Animal Welfare's approach to the control of feral cat colonies. And I was so impressed by this presentation, because I think I was relatively new in the humane field at that time, and very sensitive and tenderhearted, and the idea of trapping cats merely to put them down was overwhelming to me.

I have since learned some of the sad realities of life, but nonetheless, the idea of trapping cats, spaying and neutering them, and then returning them to responsible feeders, had great appeal, and I thought I would like to try it out at Martha's Vineyard, where my husband and I have a summer house. Usually the reputation of the island is one of a lot of happy summer people just on their yachts, sailing up and down the harbor enjoying the view. But, as in any resort, of course, there are animals abandoned at the end of the season. And this is certainly true of the Vineyard. We find that numbers of dogs, but mostly cats, will be born in the wild because people have adopted a cat for the summer or a dog for the summer, and then left without that animal, and the animal is left, to breed on its own, and the cats just seem to have a way of finding barns and finding hideaways to reproduce, and so the number grows and compounds itself.

But I wanted to try the UFAW method, and my husband and I are really very lucky, because we both worked for Pan Am Airways for a number of years, long enough to have earned lifetime travel passes. This is a great temptation to find excuses for projects all over the world. But, we decided, why not? Let's hop off to London, and look up this guy, Peter Neville. So we did, and I learned on calling the UFAW office, that Peter Neville had taken off for Thessalonika and was involved in a feral project there. But I thought, "Egad! I've come all the way to London, and there's no Peter Neville." But when I explained why I had hoped so much to see him, I was told that Peter's boss, Dr. Jenny Remfrey, was available and would be happy to talk with me. She was a little, I guess, puzzled at first, by this crazy American calling to find out about the UFAW method of cat capture, and I thought she was very clever. We turned out to be great friends, but initially, when I said, "I really want so much to talk with you" (and, since the UFAW office about an hour's drive from London), "I will come to you, or if there's any reason you might be coming to London, I'd love to treat you to dinner." There was a pause, and she said, "Well, in fact, I am coming to London to attend same sort of lecture the next day, but let's just have a drink." She was protecting herself from an absolutely awful evening it might turn out to be. But, when we met, we struck it off beautifully and, in fact, she did stay for dinner, and it happened to be my birthday, and so she participated in the surprise cake and the little stringed orchestra in the dining room playing "Memories" from "Cats," and the whole thing. And then she ran out and said, "I've something in the boot for you." Well, I now know the boot is the trunk of the car. So she went out, opened the trunk, and came back with one of my favorite birthday presents of all time, a squeezeside cage. And you just can't really do this control project quite as well without a squeezeside cage. If you haven't seen this cage before, other than the brief glimpse on the video, this is the way it works.

There is a separate panel, which you will see, since it is being passed around. The animal is trapped in a standard trap, which you did see. This panel is pulled out, and there is a plexiglass panel that rises from the trap so that it is sort of end-to-end. The cat runs from the trap, thinks that it is going to be running to freedom, but it runs through here. And once it is in this contraption, you use the squeeze panel. You see how that accordions it in so that the animal is held like that, and it's in a tight position, so that the veterinarian or technician can stick the needle in to anesthetize it and then de-flea it, check it for ear mites and, assuming it has not had food within the last five or six hours, it is then down for spaying or neutering. It's a really ingenious device, and people who have used this squeezeside cage have said it just revolutionizes the whole concept. It makes it so simple.

If you are involved in feral cat work, and you feel you would like to order one of these, they do come from England, but we've worked out an arrangement whereby they can be bought for $50 here in the States, including the cost of importing it, and actually that's not so bad when you think it simplifies the whole problem of restraining the cat. Let me give you the specific information, because if someone would like to order it, I can tell you exactly how to do that. In fact, I have a catalogue from the designer of the equipment. It's a company called MD Components. If you would like a copy, write to:

> MD Components, Hamelin House
> Rear of 211–213 High Town Road
> Luton, Bedfordshire
> LU2 0VZ, England

I then returned to the States with my new equipment, and started using it at Martha's Vineyard with a fair amount of success. But the real challenge: and the pioneering venture that I want to describe to you today was the challenge that was dropped on my lap when I visited Virgin Gorda in the British Virgin Islands, and learned that the cats at the three adjacent resorts in the area were multiplying rapidly. A few cats had been imported, actually, and this is what's so sad—to think that someone would bring in a couple of cats to kill the rats, and then start destroying the cats when they multiply. And yet it's so inevitable that they are going to multiply unless you've taken steps to prevent this. But there was one really cat-hating manager, the very one who imported the cats to begin with, who said, "This is just becoming a nuisance. We can't have it." And so he was drowning the cats his man had been catching. But the local people believe that it is very bad luck to kill a cat, and so, when he said to the local workers, "If you see any cats, into the brink," they just didn't deal with it that way. Their approach was "Out of sight, out of mind." They would take a cat, trap it and remove it to a barren island, thinking the cat would manage. But they'd neglect to think of the fact that there are dry seasons in the British Virgin Islands, months maybe when there won't be a drop of rain, and the poor cats would just die a slow and agonizing death with no water, nothing to eat. So that's worse, even, perhaps, than the drowning.

But this nasty manager—who since has seen the light, so I shouldn't call him nasty—to make his point to the locals, one day said, "Look, I mean it. I say: 'No cats'." And to make this statement strong enough, he just picked up a cat that had been trapped and hurled it live into an incinerator, and that's what set me into action.

Well, it really was divine intervention, because the very next day that I found out that the cats were becoming a tremendous nuisance, the head of the department of veterinary medicine of Tufts University happened to be on the chartered yacht in the Sound, and attended a cocktail party at one of the three resorts. Life is just amazing how these things happen that are truly more than co-incidence. One of the manager's wives had an old dog that she adored, and she asked Jim Ross, the head of vet medicine, would he please be so kind as to clean her dog's teeth, because we didn't have a veterinarian at Virgin Gorda. Dr. Ross came up to the manager's house, cleaned the dog's teeth, and because it's just such a beautiful part of the world, he said almost half-jokingly, "Any time you have a need for me, just call on my services." And the manager's wife told me this, and said, "You know, isn't this funny—your coming to me now and saying that you'd like to do something to help with this cat situation. Because today, Dr. Ross just said he's available."

When I returned to the States, I called up Tufts Veterinary School and spoke with Dr. Ross. I told him about this program of capturing and spaying and neutering, and asked him what he thought about having his senior veterinary students use this under the supervision of their professors, use it as a mini-internship. He thought it had possibilities. It took from January of last year until July, when we actually started the program. But I then had not only to find a way for Tufts to do this program, I had to raise the money. We did that through local people like Laurence Rockefeller, who owns a resort down at Virgin Gorda, and by involving people who had had experience with that area and who cared about the animal situation.

So we did do the funding. But most people who are going to give a substantial amount of money for a pioneering venture like this won't do it unless they're going to get a tax break. Then we faced the problem of "This is British territory, and how do you get a tax break on U.S. dollars?" Well, we did it through the Center for Animals at Tufts University School of Veterinary Medicine. It was using an American university, and it was a training program for the welfare of animals, and it was certainly helping the cats. It was a legitimate way to do our funding.

We also had the technicality of how American veterinarians can go into a foreign country and perform surgery. There are regulations about this, and we had to overcome those regulations by getting the local Lions Club to say they were the sponsors, and thus it went. But we did it.

It was a really exciting program, because one professor and two seniors would appear for one week of spaying and neutering all the cats that were trapped, and then they'd be replaced by the second team, and then the third week, again, another team. So that there were enough students who were really benefiting from this chance to do massive quantities of spaying and neutering to make it worthwhile from the viewpoint of the university.

I think the best thing of all that came out of this project was the awareness of the local community: the people of Gun Creek, the little local town adjacent to the three resorts. These people were very unaware, as are most people throughout the Caribbean, of compassion for animals. The locals would maybe have three dogs that they'd just sort of keep out in their yards, and they'd feed some cats that would come around, but they certainly didn't consider them house cats. For these people, who quite often worked at the three resorts, to see these doctors in their little green outfits like "MASH," performing surgery in a conference hall at a posh resort, was fascinating. They call the cats "bush cats." We call them "feral cats." They call them "bush cats," and they refer to them as being "cut" rather than "spayed." This summer, when we did our follow-up, they would tell me, "You cut my cat last year. She's been fine ever since, since you cut her." Well, no kittens.

They really entered into the spirit of the thing, and they brought their own cats. Forty-seven cats from the surrounding town, the native town, which certainly made a difference in their awareness for animals, for the future.

We accompanied this with a program of humane education. We subscribed to a copy of *Kind News* for each child. This is a publication of NAAHE,[6a] the educational branch of HSUS. Each child got his or her copy of the newspaper delivered right into the classroom and, luckily, the headmistress of the grade school was one of the exceptions who truly loved animals. She welcomed the idea of having this material for the children, and she received her copy of the magazine for teachers, *Children and Animals*. Of course, each time we go back there, we check in at the school and see how the children are responding. And there's been just a lovely change of attitudes, and that's the most satisfying thing to me. Since then, we've brought in some of our New York Humane Education Committee materials, and it's all going very, very well indeed.

Just quickly, about the follow-up—we knew that it would be an idealistic dream to do this just once, and then never again. So, this year, we went back with a smaller group from Tufts University, two professors and four students, and spent just one week. But we had Peter Neville with us this time. He went down to Virgin Gorda just this last month. He was up about 2:00 a.m. and 4:00 a.m. and 6:00 a.m., checking the traps. We found that quite a number of the cats that had been caught last year were silly enough to go back into the traps. (Maybe the quality of food was good in there.) But we could, of course, identify them by the nicked ear. Just a little of the top of the left ear was missing. So we knew that we were pretty successful. Yet we did catch 31 cats that had not been spayed or neutered amongst all three resorts. That would seem to indicate that you do have to keep doing this, because just a couple of cats producing a litter, and they in turn producing a litter, within a year's time could easily give you 31. So it was probably just a couple of cats that we had missed, but we got them, and once again, the response from the community was a very positive one. That's the pioneering venture.

**LUCY KATZ:** I'm from Long Island, and I'm facing the same problem in Franklin Square. A few weeks ago I found a colony. I'm working with somebody, but she wants to take them all home, and we can't do that. My question is, if I trap them and they are wild, will the vets take them?

**WASHBURN:** That's the problem. When they are kittens, they can be rehabilitated fairly easily. When you say: "Will the vets take them?" you mean, would they even deal with neutering and spaying them?

**KATZ:** Yes, I want to have them neutered.

**WASHBURN:** They will with a squeeze-side cage, because I found the difference of night and day at the Vineyard, (where I did this before), between just bringing in a cat in a regular carrier and having the vet say, "I don't want you to do this to me again," as the cat is climbing the screen of the window. But with this, they all would just say, "Oh, my gosh, where did you get this thing?" So it makes all the difference.

**KATZ:** My problem is immediate. By the time I get one from England, I'll have 50 kittens.

**MULLEN:** Yes, that's a very real problem, but we found a way of dealing with this that would enable a veterinarian or a veterinarian's assistant to handle the situation very safely. If they have the patience and the willingness, it's very simple to pre-tranquilize the cat through the bottom of a regular cage. It requires two people. If you have a trapped cat, you tip the cat up this way, and I have a Tomahawk trap' that's approximately this high, and a bit longer. You tip the cat up like this. The other person goes under it. It's very easy to administer an abdominal injection or a subcutaneous injection of sufficient quantity to anesthetize the cat and make it easy to handle. But I'd rather have a veterinarian address that. There's one standing behind you. Dr. Lerman?

## Control of Feral Cats

**DR. MARK LERMAN:** Yes, I thought I'd address that. The squeeze cage is really nice, but even a standard trap will do. I work with a lot of people who bring in feral cats. We do a very similar type of a thing, and they bring them in usually right in the trap, because a lot of these cats are really, really difficult, frightened animals, and a lot of the people who bring them in aren't that experienced. They love the cats, but they aren't experienced in handling, so they bring them in the trap and, usually, all we do is put it on a table, and if one person stands on one side and waves a hand or something, the cat usually backs into a corner. Then we use ketamine, which is an intramuscular anesthetic, and if you're quick (it takes a little practice), just jab it and that's it. We usually use a real low dose, half a cc or so. But that's just enough to sedate them real heavily, and then we can examine them and put them under a deeper anesthesia for surgery. And most veterinarians, if they're willing to take the time, can do it just fine, even in a regular trap.

One thing is finding veterinarians to do this. And if I could address that to some extent. I think that you have to realize that most veterinarians are very busy and aren't as concerned as a lot of the people, as all the people here are, and I think that if you approach a veterinarian in advance to set up a program and go in, make an appointment and talk with the veterinarian instead of calling up and saying: "You know, I've got five stray cats here. Could you spay them next Tuesday? Well, I'm not sure when I'm going to be able to trap them, but as soon as I trap them, can I bring them right over, because you know, you can spay them, and I'll pick them up in a few hours." If you can work out something in advance, I think most veterinarians have time, little bits and corners of time, especially in a multi-veterinarian practice, where they usually have a few veterinarians who are junior and are being paid to sit there anyway, and if you could do it in the wintertime or the time that the veterinarian is not as busy, these things can be worked out. But the thing I think is really important is to discuss it in advance, and I think you'll be able to do it, because veterinarians are really quite capable of handling these feral cats if they want to. They can use it as a nice excuse for not doing it, but we're pretty experienced, most of us, in handling these animals.

**WASHBURN:** Thank you so much, and I wish the world were peopled with vets like you.

**VOICE:** How do you deal with feline leukemia in feral cats?

**WASHBURN:** That is a point well taken, too. It *is* a threat. Luckily, for last year's project, down at Virgin Gorda, Sue Kotter was one of the doctors, and maybe you're aware that Sue Kotter was one of the pioneers in working with feline leukemia. Tests were made, and so on, and amazingly, because it is an island, leukemia has not presented itself there. But I know it is a tough thing in a normal community where you're hoping that your pet cat won't become a victim of the leukemia virus, because this other cat is strolling around in the neighborhood. It's my understanding, however, that as far as contracting leukemia, that it has to be through the saliva, direct contact, and usually over a period of time—that just one contact won't usually spread it. But there is always that question, and I think it's a valid point to raise. And yet, on the other hand, those of us who love cats would hate the idea of putting the whole colony down just on the chance that maybe one out of 99 might have the virus, so it's tough. Have you had experiences with leukemia overtaking groups that you might have been involved in trapping, or what was your thought?

**VOICE:** Well, we're just trying at this point to get this program going. I have the space, and I have a lot of people that need it on the Jersey shore. If cats have feline leukemia, what do you do with them then? Euthanize them? And there are some who feel very strongly against not euthanizing them. We need some answers, some help in that direction as to how can the group come to a moral decision on this.

**WASHBURN:** I know, because the same thing applies in shelters, too, where they test for leukemia. What do you do if they're positive? So, it's rough.

**MULLEN:** There is one very important aspect that we haven't touched on in this presentation. Many colonies of cats simply cannot be returned to the spot. People do not want to continue feeding them. The program that I work with in Ulster County has over the past three years tracked 20 volunteers' records which I examined recently. Our most active volunteer has trapped some 885 cats we've had to euthanize. In addition to those, she was able to have 107 cats spayed or neutered, and placed them. We don't place cats that aren't tested for leukemia. We pay for the tests on the ones we catch, but it's a financial burden, and it's one that you have to be willing to assume. Otherwise, you're really doing only part of the job, and possibly exacerbating a very serious problem. I don't know what the gross statistics are on the incidence of leukemia. Dr. Lerman?

**LERMAN:** It's estimated that somewhere up to 10 percent actually carry it, but it's probably more like about 3 or 4 percent. But I've seen cats in colonies where significant numbers of them have it—about every other cat has it. So that has to be a philosophical type of a question, not a medical one. Medically, the best thing would be that any cat that comes out of a colony that's feline leukemia positive should be euthanized, and medically, it's pretty clear-cut. What I see in these colonies is that the people who bring them in don't want to do that, and then we have a real problem.

**MULLEN:** Yes. I think you could say that this is an issue to which we could devote a great deal more time. Unfortunately, we're going to have to take it up at another date.

# Hell in Paradise: Vieques, Puerto Rico

**MULLEN:** Now I would like to introduce Ann Cottrell Free. Ann Cottrell Free received the Albert Schweitzer medal in 1963 from the Animal Welfare Institute. She's a freelance journalist and author, who writes beautiful articles which you might have seen from time to time in such publications as the newsletter of the Animal Welfare Institute. She's responsible for initiating some major publicity—going really public, as it were, with the problems of the decompression chamber and pointing out to the public that that was really a most inhumane method of dealing with the overpopulation problem. Euthanasia in itself is a sad, but not necessarily inhumane, method of addressing pet overpopulation. Certain methods of destroying animals can be tremendously inhumane, and Ann Cottrell Free, through her gifted writing, brought that problem to the public in a way that helped ultimately ban the decompression chamber in many states. She did that writing in a 1971 article published in *The Washingtonian* magazine, "No Room Save In the Heart." That is also the title of a book she's just recently published.

**ANN COTTRELL FREE:** The reason I am speaking here today and after AnnaBell is that my subject is an island rather close to hers, Virgin Gorda. The American flag, however, flies over our island, the island of Vieques, which is a part of Puerto Rico.

I bring you a horror story from paradise, the most beautiful island I believe I've ever seen. I've had a long life in the world of animals and suffering, and never have I seen anything like what I have experienced in Vieques. Therefore, I've spent a great deal of time since my last trip there in April, trying to do something about it, and I have hope—a great deal of hope. I believe that we can lick this situation I'm going to tell you about.

I was there when I saw dogs, puppies, dropped at the dump. And you've all heard of that before, but it's pretty horrible when the bulldozer bulldozes the puppies in front of their mothers, when the ones who have survived have no hair left on them, they're scrounging, there's no water. The big dumping ground is the "solution" to overbreeding. There's no veterinarian, no animal control, no humane society, until now. The horses run loose. They fall. They get sick. A horse was injured on the beach. Its leg was broken, it was suffering, it started floundering in the water, and they called the police. No one would shoot it, because it might be somebody's property. So somebody had some T-61, and the cook at an inn, whose father was a veterinarian, used the syringes five times into the vein of the neck to kill it. And this is just bad stuff. When there's a drought, the cattle fall and die. So, now, those are just sort of the highlights, highlights of the horrors.

Where is Vieques specifically? It's right across by plane about 25 minutes from San Juan. It is about 7 miles from the main island of Puerto Rico. It is about 5 miles wide and 21 miles long. It has a population of about 7,000 persons—most on welfare. In 1941, the United States Navy took over, because the war was on. The people were dispossessed, just thrown out, more or less, and the land bought up. The Navy owns two-thirds of the island. They have live ammunition firing. Ammunition is buried at one end of the island.

In 1978, the fishermen, who wanted to indulge their old rights of fishing in the simple, primitive way, were protesting the firing across their fishing areas, so they accosted the U.S. Navy, and out of this they got a lot of publicity. They won out on that. Then a group came along after that, really quite annoyed by what had been done to the ecology by all the firing, and they sued. So now we've got what they call the Vieques Accord, to provide economic opportunity and protect the environment.

I knew the situation was bad in Vieques. But this time, when I arrived and had a rental place, there was a mother dog that came out of the bushes, and I thought, "Oh, oh, puppies back there." So sure enough, there were, and I really couldn't abandon them because there were puppies everywhere, all over the place. And so we managed to get them over to Roosevelt Roads, the big naval base right at the end of San Juan, in a boat that goes twice a week. A navy veterinarian is there. I could use him, because my husband has a legal connection, a retired officer with privileges. It was certainly great, we could use the vet.

## Hell In Paradise: Vieques, Puerto Rico

But he used T-61, I'll have you know. Not good.

Well, in any event, I came back over and then I went to the dump to find the real source, and that's where the horror story really began. I realized that this situation just couldn't go on, because they have these puppies all over the place, and I found this woman, Rebecca Kitterman, who had tried to get a humane society going a few years ago and a plea had gone out to humane societies in this country and to the U.S. Navy. Nothing happened. "Oh, mañana"—things falling between the cracks; no money. But my contention is the following: You can put the finger, you might say, on the United States government because they have kept down the economic opportunity for the people. We must concentrate on the U.S. government, I believed.

So when I got back home in Washington, I wrote a flyer, which I sent out to humane groups, suggesting that we could get a two-pronged program going, a stop-gap one to get enough money to get a collector, to pay for a technician to do euthanizing, and start on some free spays, and to get a humane society together, so that it could be recipient of funds, to be a nucleus. Then we go to the Navy, which is our next step. It's a long-range program. The Navy, although they are responsible for this, are not thinking about giving out any grants, unless you've got a track record.

Now, and this has only been a few months, they've been having meetings. They have a humane society going there. You'll be pleased to know that the articles of incorporation have been drawn up by a lawyer, who's the only lawyer on the island, and he did it free of charge. And in the meantime, I've gotten a lot of checks and, indeed, they have too—to start paying for this extra work that I've been telling you about, euthanizing and so on. But it still is just a drop in the bucket. The next big step is to get a humane program going for about a year to establish the track record and then to show the U.S. Navy and the municipal government that this can be done. Then it can qualify for, I hope, a good grant from the Vieques Economic Adjustment Program, which would grow out of that rebellion that the fishermen had, plus the lawsuit. So that in other words, the Navy knows it has responsibility. We can go to them after we've shown we can do things.

Now, in the meantime, I went ahead anyway and applied to the Albert Schweitzer Animal Welfare Suggestion Fund. And, I'm happy to say that a small grant of about $1000, $1900 will be forthcoming. But that is only a drop in the bucket, and I'm really desperate for funds. We need to get a veterinarian down there. I feel that the spaying should be free unless you're riding up in a Cadillac, because everybody's so poor. You almost have to pay them to come. You get the birth control thing under control, and then you keep these dogs off the dump. Some of them.

The rest of the horror stories are really pretty awful: animals being set on fire over at the Marine base, a cat was put in a washing machine with Clorox. You wouldn't believe it. All I have to say, is that we want to go ahead with coping. I'd like to see them get about $30,000 in grants. I think that's how much it would cost to run a program per year with the free spaying, euthanizing, collecting, one full-time employee, that means maintenance on the little collecting house, and a truck.

I'm pleased to announce that I had been putting out this flyer, having checks come to me, and I just call it Vieques Animal Emergency Fund. The new name, legally, is Vieques Humane Society and Animal Rescue. If any of you as individuals or as societies wish to make a contribution, you could send it to the address on these flyers [4700 Jamestown Road, Bethesda, MD 20816] which is my personal address, and I would just forward it.

We *must* use some of that money for education in Spanish. The Society is getting out flyers, and they *are* getting donation cans for the stores. But we must get into the whole educational thing. I went to see the priest to ask him to do some homilies and that sort of thing. He said he would. So I believe that we should keep after this, then go to the U.S. Navy, and say, "Look, we've come this far. They can't do it on their own. Now, would you come in here and pick up the major part of this expenditure? It's your obligation." And I believe if they know that contributions have come from such substantial persons as yourselves, I think they will listen. Thank you very much.

[See Update pp. 235-236]

**MULLEN:** Thank you, Ann, for bringing a problem to our attention which most urgently needs to be addressed.

# Euthanasia: Agents Used; Public Perception; Stress on Technicians

**MULLEN:** Our next topic is, "Euthanasia: Agents Used, Public Perception, Stress on Technicians." Also covered in this topic will be remarks about a recent change in legislation concerning euthanasia.

Euthanasia, as all of you here know, is the most pervasive, it's also the most repugnant, method of controlling the numbers of animals that find their way into many of our animal shelters. It's one of the tragic realities of shelter work. Dr. Gordon Robinson is going to address this topic first. I'll give you a few of Dr. Robinson's illustrious titles: Director of the Bergh Memorial Hospital, Treasurer of the Executive Board of the New York Veterinary Medical Association, and Director of Region One of the American Veterinary Animal Hospital Association.

After Dr. Robinson, we'll hear from Ingrid Newkirk. I'll make her introduction at that point.

**DR. GORDON ROBINSON:** When I was first asked to speak here, I intended to spend most of my time discussing different agents for euthanasia. I teach anesthesia as well as being primarily a surgeon. But the recent change in New York law essentially makes a lot of this just academic. Assembly Bill A.5067-A[8] of 1987 has been passed, and I'll cover briefly some of its provisions.

It eliminates carbon monoxide automobile exhaust systems. Although pure carbon monoxide is available commercially as a bottled gas, it is expensive and dangerous to use. I would not recommend it.

Dr. Gordon Robinson

Nitrous oxide, I think, is far too expensive, and with nitrous oxide involved for anesthetic agents, and OSHA, the Occupational Safety & Health Administration, we essentially have a new term in medicine, and it's called pre-menopausal woman, meaning a woman with child-bearing potential. And we have scavenger systems in operating rooms now, so that pre-menopausal women are not exposed to nitrous oxide and the volatile anesthetics. So not only is it expensive, but essentially it's a problem when we use it in an open system that it pollutes and you have to breath it.

Carbon dioxide is very inefficient, and the volatile anesthetic agents which are the common anesthetic agents used in surgery today, like methoxyflurane, halothane or ethane isofluorane, enflurane, they're essentially just too expensive. I know the law eliminates T-61 by name, so this really essentially leaves only the barbiturates. The law allows barbiturates, or Drug Enforcement Administration Schedule II drugs. The law then goes on with a provision by which you can essentially bypass the health professions, veterinary or dentist or MD, and get Schedule II drugs—only one, though. Not all barbiturates—only one—and that's sodium pentobarbital by name, in the law. [See Editor's Note 8, last paragraph.]

Now, there are other barbiturates that can be used for euthanasia. Secobarbital is a common one. And even the thiobarbiturates, such as sodium pentothal, which you could get from your local dentist, could potentially be used, but the law does not make any provision for you to get them. It makes provision only for you to get one, and it's by name. So essentially, I think euthanasia in New York State is going to come down to sodium pentobarbital, so it doesn't make too much difference about discussing all the others.

At the end, I will help walk people through how you start to get a Schedule II drug. [See Editor's Note 8, last paragraph.] I have a copy of the solution that we make, and I'll discuss a little bit about making your own.

The second topic is public perception. Most humane societies or animal control agencies tended to de-emphasize the fact that 75 percent of animal control was euthanasia, up until the last maybe five to eight years, and now it *is* emphasized that a lot of animal control is euthanasia. And it is usually emphasized as a way to push their spay/neuter program. We emphasize, unfortunately, that we euthanize 80 percent of all the animals we control. And the Massachusetts SPCA has started a massive spay/neuter program, or is attempting to start a big one, and a lot of their publicity emphasizes that animal control is euthanasia. I think the public didn't have to think too much about animal control, what animal control meant, but as we push the overpopulation problem, we are realistic about what animal control means.

The second thing that I think, in terms of public perceptions, is that they don't like anything that has to do with a chamber. It goes back to Buchenwald and Belsen. So I think if there is any possible way you can get away from anything that has to do with a chamber, you're much better off as far as how the public perceives us in animal control.

The last thing here is stress on technicians. I think in terms of technicians, maybe a little bit differently than some other people do. There really are two types of technicians. There's what we call a handler, and this is basically an on-the-job trained kennel man. I contrast this with what in the field of animal health and veterinary medicine is known as a technician—a licensed AHT (Animal Health Technician) who has gone to two years of school.

First, our handlers who work with the euthanasia program are essentially on-the-job trained, and don't seem to object or show stress that much. What seems to be objectionable, is the fact that the work is dirty, it's dangerous and it's physically hard. If you have to euthanize 30 animals an hour—and they average 30 to 40 pounds—you bring them out, you lift them up on the euthanasia table, they're euthanized, you put them down (what we do is lay them on the floor and check them again 15 minutes later to be sure they're dead), and they're put up on a cart and then they go into the morgue. You will lift this 40-pound animal four times. It's 1,200 pounds once for 30 animals. It's 4,800 pounds in an hour. These technicians lift two and a half tons an hour. That's physically hard work. They also have to handle animals that have sarcoptic mange, ringworm, open sores, fleas, ticks, scabies, ear mites. They also have to handle animals that are either feral cats or street dogs, and they get a higher percentage of bites and scratches than the other people do. Unfortunately, if you have to do a lot of euthanasia, it's dirty, it's dangerous and it's physically demanding. Two and a half tons an hour. How long do you think a hundred-pound woman can do that? Two and a half tons an hour. That's better than you do at Jack LaLanne.

So I think, in our experience, these people must have a choice. They must have a choice as to whether to do it or not, even if it is an employment criterion. And number two, our handlers are paid more, because it is physically hard work, and they do have to euthanize a lot of animals that *do* have mange and open draining sores, and ticks and fleas, and so forth. Now, in our experience anyway, there has not been great objection, but they *must* have a choice, and I think they *must* be paid more.

Now, the other type of technician—the animal health technician. I think, as animal health technicians, with two years in a large veterinary practice, they would probably be the ideal health care supervisor for the average size shelter, that maybe can't afford to have a full time veterinarian. The problem is that they won't do it. These AHTs say, to quote one of mine: "I went to school to be on the health care team, I didn't go to school to be on the euthanasia team." And we have some well known ones, some that lecture nationally and are very well known in veterinary medicine, and I have spoken to all of them: "If you could be in charge of the shelter health program, understanding that shelter health is partly screening and vaccinating, but it also involves euthanasia, and if it was $50 or more, would you accept the job?" And they all said, "No." So in terms of technicians, probably if you could get a veterinary technician, as opposed to an on-the-job trained type technician, it would be the ideal answer. But unfortunately, I don't think most of them will accept the job. They enjoy working with the veterinarian on the "health care team," as they call it. The nurses don't go to work for coroners either.

With female technicians, you've got to be careful of the use of carbon dioxide or the use of nitrous oxide or of any volatile anesthetics. And that's such a hassle nowadays. It almost eliminates it, because there's nothing worse than having OSHA on your back.

One thing that has improved recently is, I think as far as euthanasia, is the newer methods of controlling aggressive animals. Since the availability of ketamine and Rompun (xylazine), aggressive dogs and feral cats are nowhere near the problem that they used to be if you were trying to do anesthesia 20 to 25 years ago. It's so much easier today. Tranquilizers are relatively inefficient, and acepromazine never stopped anything from biting. But ketamine and Rompun are both excellent. They are both injectable, and I think they have decreased the stress on technicians because there is less physical problem today. I will make the point about both. Neither of these drugs is a controlled drug. Most veterinarians have thought for years that ketamine was very close to angel dust, and would become a controlled drug. To give you an idea what a problem it is, how many of you remember the TV show, "60 Minutes," about the guy that used to meet girls in a bar, put something in their drink, then take them out and rape them? Remember that story? Well, that was ketamine. That was what was being used. So, if you have access to or use ketamine, even though technically it's not a DEA controlled drug, you must absolutely control it, because it has tremendous abuse potential. The same is true of Rompun, though to a lesser extent.

A general point: all euthanizing agents, in general, affect the brain. Therefore, they are mind-altering drugs; therefore, they have abuse potential. Any anesthetic agent, any euthanizing agent, has abuse potential. Be careful. Euthanizing agents now are made usually with glycol, either ethylene glycol or a propylene glycol. They are non-irritating, and therefore they can be used intraperitoneally. For little kittens and little puppies, the use of them intraperitoneally is also less stressful because it's a hell of a fight to try and give a wild tom kitten an intravenous injection of a material that's as thick as molasses.

We went into a sodium pentobarbital intravenous euthanasia program in 1979. We make our own, and I'll talk a little bit about it. For the rest, I'll cover how you go through this hassle with the DEA and how you get Schedule II and III drugs.

We make our own solution [Note 8, last paragraph] and our cost is $8.43 for a 250 cc bottle. The disadvantage today of making your own is that, to my knowledge, the only source of powdered pentobarbital, Gaines Chemical, provides a five-kilogram container, and that is the smallest. It used to be sold as one pound, and then it was sold as one kilogram, but Gaines Chemical, as far as I know, is the only known source of the powdered pentobarbital, and the smallest amount sold is a five-kilogram container. In order to make it then, you have to make 70 bottles of 250 ccs. Now, unfortunately, that's not an awful lot for us, in the biggest city in the United States. I would recommend if you would not make it in that volume, that you don't make your own. And the reason is, number one, the powder is very irritating, and I feel that I shouldn't have women working with it. The second thing is, it's so easy

Dr. Gordon Robinson.

to be missing one ounce of this stuff, and as a powder, in purified form, it has street value. So, when we open a container, that entire container will be used. We will not attempt to make one pound worth out of a five-pound bag, or one kilogram out of a five-kilogram bag. Once it's open and it's unsealed, it's too easy to be missing an ounce of it. So when I open it, it's all used at once. I would not recommend that for you, unless you share with another humane society. It is said that it's not stable, but that's questionable.

We used commercial sources of pentobarbital solution before we started making our own. Vortech Pharmaceutical, which used to be American Pharmacal, to my knowledge is the cheapest source of pentobarbital solution. They sell by individual bottle. There are some other states besides New York that will allow sodium pentobarbital to be bought by nonprofessionals, and Vortech is very aware of how to guide you through the sequence, and how to help you get your DEA number and so forth. They're very helpful that way. They are not a large pharmaceutical company. They're a pharmaceutical company specializing in euthanasia products, I think.

Another Schedule II drug is one called Sleepaway from Fort Dodge, which is pure sodium pentobarbital. Repose, from Syntex, is secobarbital. It's a good one, and it's a Schedule III.

A little bit about what is a Schedule II drug and what is a Schedule III drug and how you get through the sequence: the Drug Enforcement Administration classifies drugs from I to VI. Number I has no medical use. That's things like heroin and crack. Number II is pure drugs with high abuse potential. Number III is drugs with high abuse potential but combined with something. And IV, V and VI get less and less significant until you get down to codeine cough medicine. It's easier to get a Schedule III drug than a Schedule II drug. The Schedule III is the one that's in combination. And many times, the additive doesn't really have any pharmaceutical significance. The only reason it was put in there was to get it to be a Schedule III drug, so therefore, it's easier to sell. And there are two of the agents that I just mentioned, one called FP-3 from Vortech and one called "Repose" from Syntex, that have local anesthetic agents in them. One has lidocaine, the other has dibucaine. I don't think they'd have any effect whatsoever, except they make it a Schedule III drug, the idea being if you combine it with something, it is harder to be extracted out and then be pure again and then have street value again. Most people couldn't separate the lidocaine from the pentobarb, and since lidocaine does have some cardiac toxicity, they are less likely to abuse it.

To get Schedule II and III drugs, you must be registered with the State, and in most cases, it's the Board of Pharmacy. In New York State, it happens to be the Department of Health. The law that was passed specifically says that the Department of Health needs to license you for pentobarbital, period. [See Note 8, last paragraph.] And then you must also be licensed with the federal government, and that's the Drug Enforcement Administration of the United States Department of Justice (the DEA). When you register with the DEA, you get a controlled substances registration certificate. And you also apply for a booklet of forms called the DEA Form 222.

Now, to get a Schedule II drug, you must submit a DEA form 222 with every single order. To get a Schedule III drug, all you have to do is send in your yearly renewal certificate to show them that you have renewed your narcotics license, called "controlled substance license" today. So, therefore, it is much easier to get a Schedule III drug than it is to get a Schedule II. The Schedule III does not need a form every time. They must just have on file a license for you, saying that you are currently registered, and from then on, they just fill it. This is why the companies have added things like local anesthetics—just to get it to be a Schedule III drug. Vortech has a Schedule II and schedule III: Fatal-Plus is their Schedule II drug. They have one called FP-3, which is Fatal-Plus with an additive which makes it Schedule III, which you can get more easily.

**MULLEN:** Perhaps we should take questions now.

**ANN JOLY:** I'm with the Massachusetts SPCA, and I'm the assistant manager of the Boston shelter. Just a quick thing about stress with the technicians: My technicians have never, ever, said one comment about anything physical, about lifting animals or getting mange, or anything else. So I think that's the furthest thought in their mind and the furthest in my mind. I, as the assistant manager, participate in approximately 96 percent of the euthanasia on the site, and there has never, ever been a concern. I think by far the main mental stress is worry about the animal, if he's having a peaceful euthanasia. These are animals that we get attached to, my technicians and my ward attendants get attached to, and I think it's a huge problem, but not a problem with my back during euthanasia. And if I ever heard anyone complain about physical conditions or getting urine on them or feces or mange—I don't want them in the room.

**MULLEN:** I would like to comment on that myself. I think they are *both* major sources of stress. My experience has been much more like yours, in that I've heard few complaints from euthanasia technicians in the shelters that I have direct experience with, about the actual physical stress. But I know personally whereof Dr. Robinson speaks. It's just that I don't get any complaints about that. I know there is a great deal of stress—it's very real. But it's the mental, the psychological stress that causes more burn-out than anything else, when having to deal with this issue of euthanasia as the form of animal control in the shelter.

**DEBRA FELIZIANI:** I would like to concur and say, having been involved in animal euthanasia in the past, that euthanasia *is* very stressful—emotionally stressful as well as physically. I flinch at the idea of having one technician, and I flinch at the idea of 30 euthanasias in an hour. I think that there is a problem in management if that's the case. My own experience in euthanasia has led me to want to make a different kind of contribution, and it's led me to the need for education and humane law enforcement and other types of issues that have become personally important to me. I take great exception to thinking that there are people who are doing this on a daily basis who have become hardened to the idea. I don't think that that's why people get into animal control, and if it is, then there needs to be reassessment of staff evaluation and training.

**VOICE:** I work at Bide-A-Wee. I did work for an animal control organization in New York City, handling strays, for a year, and I can tell you from personal experience about burnout. Euthanasia is stressful emotionally, probably more than anything else, if you really care about animals. I'm not saying to hire a person that doesn't care as much, but I would like to see in shelters some kind of counseling groups, rap sessions, whatever, for people to whom the whole concept of working

## Euthanasia: Agents Used; Public Perception; Stress on Technicians

in an animal organization is stressful, whether it's a kennel person, a handler, a clerk. One handler doing it straight every day could very easily go through 100 animals a day being put to sleep, and it's just a lot for one person to deal with. What I would strongly recommend is some kind of counseling for kennel people in animal control situations in New York City.

**MULLEN:** That's an excellent suggestion. Those of you who have attended Humane Society of the United States workshops, know that this is an issue that HSUS has put a tremendous amount of emphasis on, within the past several years. There's a great deal of literature that emanates from HSUS and other organizations about euthanasia technicians' stress and burnout. It is extremely important for shelter technicians to be able to talk out their problems with people who understand what they're facing, because it's a special form of stress that few people can possibly understand—often, not even euthanasia technicians' own family members and sometimes their co-workers who aren't involved in the performance of euthanasia.

**PAMELA MARSEN:** I'm humane educator for the Bergen County Animal Shelter Society, and I think it's interesting that the last three people are shelter employees. It's a hell of a thing to have to deal with, whether you're at the ASPCA or any facility, having to put down animals. I just wanted to ask Dr. Robinson: I know some shelters, including the ASPCA, use animals that are slated for euthanasia as blood donors for their hospitals or other hospitals, and I have never understood: Are they anesthetized, is all the blood drained out of them, is there any pain involved? And I just wish, since you're here, that you could clarify that.

**ROBINSON:** We almost never do this to animals. We are on the borderline of Harlem and we're in the trauma capital of the world, and we get a tremendous number of animals hit by cars. And if we are going to use blood this way, we will get an animal scheduled for euthanasia. If there is none scheduled, then we just go without it. The animal is put under anesthesia, with Surital or some other anesthetic agent. We take one or possibly two units of blood from a dog, while it is under anesthesia, then we euthanize it. The amount of anesthetic agent that is in the blood is relatively insignificant when it becomes diluted in the recipient. Nobody bleeds out an animal, because as you get to the second half of the blood, it becomes toxic, and you should just never use it. It's worse than not transfusing the animal at all.

**ROBIN REMICK** (Tompkins County SPCA): I'm not clear about the Schedule III drug, FP–3. Because it's not pure sodium pentobarbital, under the new legislation, is that still acceptable?

**ROBINSON:** The law eliminates certain methods; it doesn't say any specific method has to be used. It eliminates some by name, such as carbon monoxide from exhaust, and T-61. The reason people think that it says you have to use sodium pentobarbital is that the second half of the law allows you a way to get this, when previously you could get it only through a veterinarian. So people think the law says you have to use sodium pentobarbital. It doesn't, really. You could use Surital, you could use sodium pentothal, you could use any anesthetic agent used today. But it allows you to get one of these—which is the most common one, obviously. Yes, you can use FP-3.

**MULLEN:** If there are no further questions of Dr. Robinson, I'd like to introduce our next speaker.

## Euthanasia: Agents Used; Public Perception; Stress on Technicians

**MULLEN:** I think you may have been somewhat surprised to see the words "Ingrid Newkirk" behind the subject of euthanasia. Ingrid Newkirk is so often associated with saving the lives of animals and with being a champion of animal rights, that many people in this room might expect Ingrid to step up to this podium and make a plea to save the lives of all the animals that make their way into the local animal shelter. I think you're going to be very surprised if you think that Ingrid has that type of perspective. She has a background in animal control that many of you possibly don't know about. She served for over 13 years in the animal control field. I myself was completely unaware of Ingrid's position on euthanasia as performed in animal shelters, until very recently, when I came across an item that is one of the most perceptive and sensitive pieces I've ever read on the subject of euthanasia as performed in animal shelters. It's in your folder. It's an *Animals' Agenda* interview with Ingrid, entitled "Commitment and Compassion in Action"(Agenda Magazine; July/Aug 1983, pp 4-5). I'd like to call your attention to the portion of the interview which starts with Ingrid's comments on the subject of euthanasia. You may think it is a crime to kill animals by the millions in shelters, and you may believe that those animals' lives should instead be saved. We wish their lives *could* be saved; we know they can't. Ingrid eloquently expresses *why* they can't, and also why it's essential for a person who cares deeply about animals, to be the one who kills them. If such a person isn't there to do that, the animals won't be euthanized; they'll only be exterminated.

**INGRID NEWKIRK:** As Dr. Robinson said, in New York, you are lucky to have fewer options now than other states have. We hear about cases on an individual basis from across the country, of people suddenly discovering that the dog warden in their town is still shooting dogs, or they're using gaseous compounds that are unfiltered, that are not cooled, that are coming right out of an engine or through some other means. Around the world, animal welfare issues are still one of the paramount problems that we have to solve. We in the animal rights movement owe a great debt, I believe, to people in animal welfare who started the whole ball rolling—getting people to think first, perhaps, about the carriage horses on the streets (although we're still thinking about that), the beasts of burden, the problems of dogs and cats who became surplus because of society acquiring them frivolously, and throwing them away, and so on. And we still

## Ingrid Newkirk

haven't resolved those situations, as everybody in this room well knows.

In Israel, strychnine is still being used as a means of killing—I can't say, "euthanizing"—but killing surplus stray animals. There is a wonderful organization, a new fledgling organization called "CHAI" (Concern for Helping Animals in Israel).[9] It's trying to do something about that.

In my home country, in England, they still use the electro*than*ator or elec*tro*thanator at the RSPCA—and that's a disgrace. It's where electrodes are applied to the dog's ears and to a hind leg, and they're made to stand in a small basin of water, and then the switch is pulled.

I still get calls from people in Tennessee and Texas and places, who say, "Our dog wardens still shoot 'em from the window of the truck," and people are mounting campaigns to stop that. The criteria for euthanasia are listed, I think pretty well, in a pamphlet called *Review of Literature on the Use of T-61 as an Euthanasic Agent*, and even though this applies to T-61, (and I would like to talk about T-61, even though it's now banned in New York), it spells out what should be the criteria for any agent, and perhaps the first five criteria are the ones that really should be absolute:

**1.** It should be painless.

**2.** It should cause unconsciousness instantaneously, and death within minutes.

**3.** It should not cause undue anxiety, alarm, fear, panic, struggling, vocalization, muscle spasms or clinical signs of autonomic activation before unconsciousness, such as convulsions.

**4.** It should *always* cause death when properly used. (They all have their stories of the dogs waking up in the refrigerator room.)

**5.** It should be *safe* for the properly trained person to use.

**6.** It should be *easy* for the properly trained person to use.

**7.** It should not be a drug subject to abuse in human beings.

**8.** It should be aesthetically unobjectionable. (That's always nice.)

**9.** It should be practical to use for the particular kind of animal to be killed.

10. It should not create a problem of sanitation or environmental contamination.

11. It should not cause tissue changes which would alter the postmortem examination or clinical results.

12. It should be economical.

The reason I'd like to talk about T-61 a bit, is that even though you've won your battle here, there are several reasons to think long and hard about the good reasons for putting the dirt on its grave. One is that other people in other states may ask you what you feel about it. It was a long battle here, as I understand it, to get T-61 outlawed, to get this compound off the market. And I suppose that's hard to do, as with anything where a company has invested millions of dollars in something that they want to reap more dollars from.

The fact of the matter, when it came to light, was that T-61 was never adequately tested before it was marketed. There was one main study that was based on a few dogs being given T-61, and a few dogs being given sodium pentobarbital. In the end, it was determined that the study was funded by the T-61 manufacturer, and the sodium pentobarbital that was administered was not strong enough and was administered in an improper dose. The dose was too low. Some that were given pentobarbital recovered. And that basis was used, as I understand it, by the American Veterinary Medical Association to put T-61 higher on their list than sodium pentobarbital. But there are many problems with T-61.

When I first came into sheltering, I had to use T-61 because nothing else was available. The problem was not being able to find a veterinarian who would go out on a limb and sign a permit for me to get sodium pentobarbital. And that's a common problem. I think it's a laziness problem more than anything else, because when you go and meet with the Drug Enforcement Administration agent, and with the state agent, if there is one that you have to meet with here, the record keeping requirements are very plain, very simple. Lots of people do it. The security requirements are very straightforward. And a caring veterinarian, knowing that it would be preferable for the animals to have sodium pentobarbital used, could easily take the time. He or she is not going to give it to a stranger, anyway. They have to have some relationship with this client. They can sit down with them and say, "All right, whatever you're using now, let's keep a log for the next two weeks, and you come back every week and show me your log, and let's see if you're doing it right, and let me come with you to your facility and see how you're going to keep it," and so on. They could *do* that.

I'm delighted to hear that you can get your own permits in shelters here, now. That's a wonderful, wonderful advance. I think vets have been afraid. One can understand being afraid to lose one's narcotic license, but I do think it's being carried too far at the expense of lots of animals. Animals on T-61 *do* cry out on occasion, which is too often. They have agonal spasms on occasion, which is too often. If it goes extravenously, it stings, it hurts, and it introduces an element of panic that needn't be there. If you have a better agent available, and sodium pentobarbital is by far a better agent, then it's your obligation, I believe, to use it. Perhaps it is a little more trouble, perhaps it is a little bit more worry, but your primary obligation has always been to the animal, and so sodium pentobarbital has to remain the drug of choice, even in an area where the option exists.

If anyone has any doubts about the findings on T-61, there's a very good report in the Institute for the Study of Animal Problems folio produced by the Humane Society of the United States. And Dr. Michael Fox is very firm in his condemnation of it. I called him just before I came up and said, "Have you had any reason to rethink any of your comments condemning T-61?" And he said, "No." He said the drug manufacturer came to promote T-61, and on each occasion they failed to make their case.

Stress on technicians? I have to agree with the people who spoke at the microphone, that there's enormous stress. Yes, it's dirty, and yes, it's grimy, and yes, it's hard work, and yes, it's heavy lifting, but the hardest thing is that you have to talk to yourself about what you're doing. The biggest part of this is, it shouldn't be that animals have to die, and we all know that, if we care about them. But the fact is that they *do* have to die, unless we can get together, abandon this talk on euthanasia and sit down and decide what we can do to place another 20 million animals a year, right now. Twenty million, right now, in the next 12 months. And if we can't do that, then euthanasia is a reality that we have to deal with. And the most important thing is that the people who care and who hurt when they do it, must be the people who do it. We cannot have people who don't care, who want more money, who want special considerations, who find the job distasteful for other reasons, being given the responsibility to kill. I know I make it sound awfully easy. I've run animal shelters, and I know it isn't that easy to find the people. But the thing is, management has got to look for them, and they've got to throw out very quickly anybody who doesn't meet that caring criterion.

I advocate very strongly that there should be some sort of dealing with stress—mental stress, emotional stress—workshops for people who work in shelters. Aid is necessary, not only for the technicians who work in euthanasia, but for the road crews, too, because what they see on the roads and what they see in cruelty cases is something that—unless they're hard, and have steel in their hearts—has to prey on their minds. I think the quality that you want in a technician is what I call "strong kindness," because you have to steel yourself somewhat to what you're doing. If you didn't steel yourself, you'd break your heart. Because you care. And the thing is, the other danger that you wrestle with, is will you become inured to the suffering of animals.

I know it's a peculiar analogy, but if you watched the Gennarelli film, "Unnecessary Fuss," there are some experimenters who are working with a baboon, and they're laughing, and they're cracking jokes and they're making fun. And those of us who work in the anti-vivisection movement find it very easy to condemn these people. But it's very easy for things to happen in the euthanasia room that are unthinking, when you're in a framework that allows you to proceed with the job at hand. And so everyone who works there must be vigilant enough to say, "Would I mind if there was a hidden camera in here right now? Is there anything I should be ashamed of in my euthanasia room? Is my attitude good right now? Are these other people's attitudes good? Is it comfortable as it can be? Is it clean? Am I rushing?" Not just because I want to go home and have my dinner on time, that's not a good enough reason. I've got to slow down and do my job and do it well and do it properly and do it thoughtfully and do it caringly, because just as those experimenters didn't remember that that baboon was watching them and thinking about them, and aware of their actions, so in the euthanasia room, it's that animal's last moments in life, and we are the guardians of those last moments, and it's incredibly important that we always try to keep the balance there, and always try to remember.

I don't think life is a great gift. Not always. You talk to people about factory farms, mink ranches, and people say, "Oh, but if we didn't create these animals, then they wouldn't have any life at all." And it reminds me of John Bryant in "Fettered Kingdoms." He wrote a wonderful piece, in which he said that if misery were a toxic gas, the stench that would permeate the air—from every little miserable crate and barrel and chain in a backyard, from all the compounds where animals are suffering, from all the road accidents—would cover the face of the earth. And that's what we deal with in shelters. Either you help the animals who are experiencing that, who look up at us (and maybe it's easier when they're crumpled and they're broken), and they ask us, "Give us the only thing you can give us." Or you let it go, you play God in another way, by saying, "I wash my hands, I won't touch this. It's bad for my karma," or whatever. Let somebody else do the dirty work, or let's adopt this animal to a substandard home, or let the animal go to a guard dog company, or let the animal go to research, or let the animal stay on the street. If you play God by giving the animal a substandard life, then all you're doing is prolonging the situation where you see that crumpled, broken animal in front of you, or you don't, and the animal dies by himself some way.

Public perception is awfully hard to deal with. I know if you stand at the counter in the animal shelter, you could go mad. You listen to this steady stream of people coming in, and they have Fido or Thibald, and they're passing him across the counter, and you hear every ridiculous excuse for turning an animal in, for failing to find or even exert any effort to find another home for them, that anyone could ever imagine. The incredible, disposable pet.

And then of course, there are the car accident victims or the victims of the losing battles with starvation and mange and internal parasites and so on. (When we talk about handling these animals, I've handled those animals before I got into animal rights and away from animal welfare, for over a decade, and yes, you get mange periodically, and yes, you get covered with things which you have to wash out, but what they're suffering is very real. It's not the passing thing that we fancy.) As the animals come in, they're stabbed, they're shot, they're beaten, they're burned, their chains are eating into their necks, and everyone says the same thing. Everyone says, "You aren't going to kill him, are you?" And what else can you do?

In a way, it's heartening, if you think about it. Here these people are, they've done everything wrong, and the animal victim is coming in your direction across the counter, and they have some grain of what really could be considered decency or a glimmer of understanding. Maybe it's something we should take heart from. Miracle of miracles, they actually thought, "You're not going to do something bad to him, are you?" And in a way, that's rather nice, and one shouldn't take it personally, one shouldn't get one's ego involved in it and think: "Are you condemning me? Look what *you* did." They obviously haven't a clue what they've done, and they haven't a clue about the volume or the options or anything else. They are children in the garden of life, when it comes to animals. They haven't been out on the truck. They haven't been out with a cruelty officer. They haven't worked in the kennel. They haven't listened to the excuses, day in and day out. So if anything, we should feel: "Poor, uneducated slob, I'll take this problem, and I'll do the best I can."

I think that it's wonderful that there is a new honesty now in shelters. I remember the first shelter I went to work for, we were taught, "Don't ever say you're going to kill them." And so that put an extra stress on you. You not only have to do it, you have to lie about it, because it's so horrible. And now I think we have other ways that we break the news and ask them their options. At PETA, we put out a pamphlet. People call us, thinking we're a shelter, and say, "What big farm do you have? You people like animals, don't you? You wouldn't kill them." And we have a pamphlet that's called *The Guide to the Sale or Giveaway of a Companion Animal*. They say, "This isn't easy." You're going to have to go through these steps. You're going to have to actually make an effort. Do you *have* to move? (Absolutely bizarre thought.) Do you realize that's a family member of yours and he's always going to think: "What did I do wrong?" You know, "What happened? Did I do something awful? Why am I here?" And they must start down that list if they can, and if they don't start down the list, they realize they failed. We didn't fail, we're cleaning up after a society that is born and lives in absolute ignorance, and the animals suffer for it.

It's like child abuse. You know there's child abuse, but how many of us actually know one abused child? How many of us know the enormity of it? You go out with a social caseworker, and then you learn. It's there. My God, it's around the corner, and it's in the next block. The enormous numbers that shelters are dealing with make it impossible to conduct their operations unless they either kill or refuse entry to a certain number of animals.

It is possible, of course, to have a no-kill shelter, and it's a wonderful thing, but it does mean that someone else is doing the killing, because you cannot have a no-kill shelter that will deal with the world's problems, unless you can get the capacity to warehouse another 20 million animals a year in these United States.

If animals can't be turned loose, no decent homes exist to accommodate them, they must be killed. I think we have to deal, we have to struggle with our religious teachings. We have to struggle with the things we were taught as children, all the stuff that there is in our brain about death and about killing. But the option is not there. There is only one decent thing to do. For many shelters that I've been in, the last moments in the hands of that technician, in the hands of the handler, have been the nicest moments that the animal has ever had. *You can see it in their eyes.* Caring people who work in shelters have an obligation not only to deal with their own stress, but with the stress of those animals, because they should never know where they're going.[9a]

I have the luxury of not euthanizing animals any more. But I would say that those of us who don't owe an enormous debt of gratitude to those of you who do.

**MULLEN:** Thank you, Ingrid.

*Left:* **Dr. Marjorie Anchel,** President, NYSHA; **Frank Rogers,** Chairman of the Board, NYSHA; and **NYS Assemblyman Morton Hillman.**

*Below:* **Dr. John Kullberg,** President, ASPCA.

*Above:* **Christine Stevens,** President, Animal Welfare Institute —Keynote speaker at conference.

*Right:* **Martin Kurtz,** Director, Bureau of Animal Affairs, NYC Dept. of Health

*The titles in the captions are those held by the conference participants as of September, 1987*

*Above:* **Margaret Geraghty,** Executive Director, Chemung County SPCA, Elmira, NY

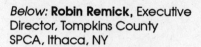

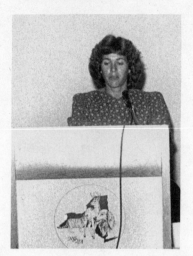

**Gretchen Wyler,** Vice-Chairman, Fund for Animals; looking on, **Margaret Geraghty**

**Barbara Cassidy,** Director, Animal Sheltering and Control, HSUS; with **Samantha Mullen,** Public Affairs & Programs Administrator, NYSHA

**AnnaBell Washburn,** President, Pet Adoption & Welfare Service, Martha's Vineyard, MA

**Ingrid Newkirk,** National Director, PETA

**Ann Cottrell Free,** award winning author and environmentalist

**Gordon Robinson,** V.M.D., Vice-President, Veterinary Services, ASPCA

**Kathleen Young,** Manager, Bide-A-Wee Home Association Animal Shelter, Westhampton, NY

**Dr. John McArdle,** Scientific Director, New England Anti-Vivisection Society, Boston; at left, **Patricia Valusek**

*Right:* **Dr. Robert Case,** Past President, New York State Veterinary Medical Association

*Above:* **Dr. Arthur Baeder,** Secretary, New Jersey Veterinary Medical Society

*Right:* **Dr. Leo Lieberman,** a strong proponent of early-age spay/neuter of companion animals

**Dr. Wolfgang Jöchle,** specialist in the use of chemosterilants to control population of domestic animals

**Ronald Scott,** Director, Argus Archives, New York City, and Garrison, NY

**Sheri Trainer,** Director, Educational Services, Bide-A-Wee Home Association

**Dr. David Samuels,** veterinarian practicing in New Paltz, NY

**Dr. Tom Regan,** Professor of Philosophy, North Carolina State University, Raleigh

**Dr. Murry Cohen,** psychiatrist practicing in New York City

**Elinor Molbegott, Esq.,** Vice-President for Legislation, ASPCA

**Assemblyman Maurice Hinchey,** honored for sponsoring mandatory spay/neuter bill, and for introducing a resolution in support of the goals of NYSHA's conference on companion animal overpopulation; with **Samantha Mullen,** NYSHA

**Grethen Wyler,** with **Senator Frank Padavan,** recipient of NYSHA award for sponsoring legislation which resulted in stronger protection of New York State animals against pound seizure, and in making more humane euthanasia procedures available for use in animal shelters

# DAY

# TWO

# Responsibilities of Shelters

**MULLEN:** Patricia Valusek has been a very generous volunteer for the New York State Humane Association, and she's graciously agreed to introduce our speakers for the first session.

**PATRICIA VALUSEK:** Good morning. We have three members on this panel, "Responsibilities of Shelters": Barbara Cassidy, Director of Animal Sheltering and Control of the Humane Society of the United States; John Kullberg, President of the ASPCA, an organization based in New York City that handles about 80,000 animals each year in its shelters, investigates and prosecutes cruelty cases, and promotes animal protective legislation on all levels; and Kathleen Young, Manager of the Westhampton Shelter of Bide-A-Wee Home Association.

**BARBARA CASSIDY:** Thank you, and good morning. As director of animal sheltering for the Humane Society of the United States, I visit about 25 to 30 shelters a year, so the responsibilities and standards of shelters is a subject that's very near and dear to my heart, and one that I've worked for diligently, for almost 17 years now.

The overall responsibility of the shelter is to the animals in its care, and the comments that I'm going to make, I believe, apply to both the private non-profit animal shelter as well as the municipal facility. I really don't think there's a distinction at all in the responsibility of the shelter itself.

We talked yesterday about several major responsibilities, one of which is the responsible adoption program, the program that places animals only in lifelong homes where they receive proper care and treatment. To do that, it is the responsibility of the shelter to have a program that includes established policies that the shelter has in place. For example:

**1.** The use of an adoption application is part of the pre-screening process to find suitable owners. Certainly, it is the responsibility of the shelter to refuse the adoption of animals to unsuitable owners.

**2.** Pre–adoption counseling, or what I call the mini-humane education program, is where the potential adopter gets information on responsible pet ownership, licensing, the companion animal's needs and the local ordinance.

**3.** The contract that includes mandatory sterilization of all adopted animals.

We do have to call it mandatory *sterilization*, because we don't think it has to be just the females. We must also be sterilizing the males. It's a requirement that some shelters haven't enacted yet. The adoption should not simply be a process where people can come into the shelter, look around, pick out an animal and sign it out. The HSUS does have guidelines for responsible adoptions, and we have guidelines for almost every other aspect of shelter operation.

It is also the responsibility of an animal shelter to conduct an ongoing humane education program. How many shelters actually budget for an ongoing program? I personally see too many shelters that concern themselves only with the business of sheltering. Humane education cannot be effective as a hit-or-miss, or "one-shot" visit to a school program. It must be on a regular basis, almost a daily responsibility of the shelter in the community. You must make the community aware of the existence and function of your agency, inform the public about the nature and scope of animal-related problems and solutions.

The shelter needs to be part of the solution, provide information necessary to motivate responsible behavior towards animals by the people in the community. And the key to the agency's ability to educate the public does depend largely on how well educated the facility is itself—the education of the staff and the board of directors, if it is a non-profit, or the legislature or government agency that supervises it, if it is a municipal animal control facility.

It is imperative that you budget adequately for a humane education program, that you have a humane educator on staff, whether paid or volunteer, but a person that creates and administers the humane education program in the community. And by that, I do not mean pet visitation programs that so many shelters do, or pet therapy programs. That is not humane education, and it is not the job of the humane educator to cart animals around to schools or nursing homes. Pet therapy is a specialized area that should be administered by qualified professionals. Regrettably, it is too often something that is dumped on the humane educator. It's not humane education.

It is also the responsibility of the shelter to provide the best care possible for the animals that enter its door. This is an area of very particular concern to me, and an area where I know that we still need to do a lot of work.

Those who undertake the sheltering of animals are morally and legally obligated to meet basic standards for animal care. Animals must be housed in such a way as to minimize stress for every species. You must provide special care and have proper accommodations for infants, elderly, sick and injured. The shelter must provide protection from the elements and provide appropriate heat and ventilation. The shelter must be cleaned and disinfected daily and designed overall for the safety and comfort of the animals that are in the shelter. That responsibility also extends to the internal record-keeping and accountability for every single animal that comes through the door of the shelter, whether that accountability is a returned-to-owner, an adoption, or a humane death.

It is the responsibility of the shelter to provide the most humane death possible. Ingrid Newkirk spoke so eloquently about euthanasia yesterday, that I don't think I need to say much more, other than that is not just the responsibility but an obligation for shelters to provide the most humane death possible, by skilled, compassionate personnel, whenever necessary. And "whenever necessary" is an important thing to keep in mind. You cannot compromise the adoption standards just to get an animal any home, and you cannot keep them on and on, indefinitely, in a cage, deprived of companionship and love.

The shelter is responsible for the ultimate fate of every animal that comes through its doors and it's responsible for the quality of their life and the quality of their death and, ultimately, for the quality of life of all the animals in the community in which it serves.

Dr. John Kullberg

**VALUSEK:** Thank you. Dr. Kullberg will speak next.

**DR. JOHN KULLBERG:** I'm going to start this morning by taking advantage of you a little bit with some retrospective comments, and forgive me if these aren't exactly on the topic. I want to speak a little bit about what Dr. Robinson had to say yesterday. I just want to put some perspective on his comments. First of all, he is a veterinary orthopedic surgeon, and a spay/neuter surgeon *extraordinaire*. He is probably among the top ten veterinarians in the country for performing spay/neuter operations. In addition, he is also the chief administrator of one of the world's largest animal hospitals—the ASPCA Bergh Memorial Hospital—which, by the way, is celebrating its 75th anniversary this year. He is heavily involved with the treatment of traumatized animals. Approximately eight to nine of every ten animals that come to our Bergh Hospital are sick or injured animals, whereas in normal veterinary practices, it's about one or two out of ten.

Dr. Robinson does not regularly interface with our shelter humane technicians and animal care supervisors. They are under the supervision of a veterinarian by the name of Dr. Lloyd Tate. Stress exists among our animal care supervisors and among our kennel staff, a lot of stress. Forty thousand animals a year go through each of our New York City shelters. Also, it should be understandable to all of you that our handlers are very concerned about mange and ringworm. They get it. And about bites and defecation. They are frequently bitten and defecated upon. But they also must be absolutely concerned about the animals they are privileged to take care of.

With regard to Christine Stevens' comments last night: we must find a better way to sterilize animals than spaying and neutering, and it is an amazing fact that we haven't.[9b] When I advocated spaying and neutering just last week at the "Action for Life" conference at Harvard University, there was an individual in the audience who, as soon as I said: "spay/neuter," uttered an expletive, stood up and walked out. He probably is also a dedicated "purist." He was against the basic procedures involved in spaying and neutering. Think about it in anthropomorphic terms. The only thing we have to offer these animals by way of a sterilization procedure is either a hysterectomy or castration. But until a better way is found, spay/neuter is the answer, and it is a major shelter responsibility to incorporate a spay/neuter clinic into its operation or

## Responsibilities of Shelters

arrange for spaying and neutering with local veterinarians.

But now, back to the topic I have been asked to speak on this morning: animal shelter responsibilities. Presuming that my good companions on this panel will focus in some depth on the issues involving the appropriate standards for retrieval, acceptance, veterinary and housing care, lost and found and adoption programs, and, as a last resort, euthanasia, I will focus my comments on the animal shelter's role in education and legislation.

Given the volume of animals that many shelters receive (the ASPCA, for example, received some 80,000 unwanted or lost animals in its two New York City shelters last year), it might be useful to consider first, the sources of animals brought to ASPCA-type animal shelters, and the assortment of animals received.

For the purposes of this discussion, I'm defining an ASPCA-type animal shelter as one which takes any and all animals brought by the public, and therefore does not limit its wards to those that might be adoptable, although there certainly is a place for adoption-only shelters.

In addition to unwanted owned animals, or lost or abandoned stray animals, shelters also receive animals seized by law enforcement authorities during investigations. At the ASPCA, these authorities also include our own peace officers.

Some of these seized animals are taken from abandoned buildings; others, by the police and also by the fire department, by housing inspectors. Others are taken during law enforcement raids on sites where animals are used for criminal purposes, such as cock fights, dog fights and animal sacrifice rituals, all of which are very popular here in New York. Other animals are held while their owners are jailed or seriously ill.

When I returned to my office yesterday, I had a very fine letter from an individual working at the Fund for Animals saying, "By the way, what are you going to do with the animals that are owned by the homeless humans that the mayor is rounding up and putting in hospitals and other places?" That's another source of animals we take in. Simply put, ASPCA-type shelters never turn away any animals.

ASPCA-type shelters also typically have an ambulance rescue

division. The ASPCA, for example, retrieved some 30,000 stray and/or injured animals from New York City streets, parks, abandoned buildings, and private residences in 1986.

Given the variety of sources of animals coming to an animal shelter, the shelter must be prepared to provide appropriate care for these animals. When an animal shelter is not equipped or for whatever other reasons cannot care for certain types of exotic animals, for example, it needs to establish liaisons with others in surrounding communities who can. Last night, I had a letter waiting for me requesting funds to help Lifeline for Wildlife[10], a superb wildlife rehabilitation center north of the George Washington Bridge in New York State. Lifeline does a lot of very important work for us, because you find in New York City that dealing with captive wildlife is very difficult. If your shelter doesn't have outreaches, such as Lifeline, you should look for them, and support them.

The assortment within the 80,000 animals received by the ASPCA's two New York City shelters in 1986 included, of course, cats and dogs, but also a variety of wild, as well as domesticated birds, horses, ferrets, gerbils, hamsters, white mice rabbits, skunks, raccoons, lizards, boa constrictors and other snakes, and tarantulas. And this list is far from complete. It's very fashionable in New York City to have these exotic pets.

When people decide they're going to move or they're evicted from their apartments, they sometimes leave these exotics, including the tarantulas, in the apartment. If we added in the 50,000 New York City animals cared for in ASPCA veterinary hospitals and spay/neuter clinics, and at the ASPCA's J.F. Kennedy airport shelter, the list of animal species would be greater still. The 130,000 animals that the ASPCA cared for directly in New York City in 1986, by the way, was the greatest volume of animals directly cared for by any humane society or government agency in America, and that includes the animal control program run by the City of Los Angeles.

And while I'm editorializing, I might add that while the ASPCA's 13 rescue ambulances are far fewer than the 51 ambulances in the City of Los Angeles, we rescue by ambulance twice as many animals as they do, and our adoption rate is 33 percent higher,[4] thanks in large part to the more than 90 volunteers assisting our efforts. We also license more dogs and

## Responsibilities of Shelters

spay/neuter almost as many animals as the city-funded projects in Los Angeles do. And we have some 125 New York City veterinarians participating in an ASPCA spay/neuter subsidy program. Think how much more, to make the point made yesterday, in this city of eight million people, we could do, if we could get the kind of government support that we should have in the city of New York and that is available in Los Angeles.

The ASPCA is a private humane society and does not, contrary to rumor, want to get out of the pound business. What the ASPCA *does* want, is for the City to shoulder more of the extraordinary costs of running city-requested animal care and control programs, a significant portion of which costs are now borne by the ASPCA. And we're taking steps to see that the City soon does.

But back to the central topic of this panel. If all that a shelter did was to care for animals brought to it, and recirculate as many of these animals as possible through responsible—very responsible adoption programs—the shelter would still be doing relatively little to ameliorate the problems that bring so many unwanted animals to shelters in the first place. What the shelter must do, in addition, is to educate the community on the causes of the dilemmas the shelter daily faces in coping with these quantities of unwanted animals, and, working with various community groups and individuals, do everything possible to set in motion legislative and other activities that expeditiously attack the causes of the pet overpopulation problem, which are greed, indifference and other sordid, wasteful and inhumane attitudes of a materialistic, self-centered and selfish society. Our biggest problem is to do effective battle with that inherent American attitude that treats animals like materialistic commodities.

By what right do we incorporate the lives of companion animals into our own? By no right. "I've been had" for us is a pejorative term, but that's what most companion animals would say if they could speak. Some here today would insist that having companion animals is ethically wrong, a remnant of slavery. For example, we, the *masters*, own others. We demand *obedience*, even *shackle* our *commodities* at times and give rewards based on *good* behavior. Bad behavior often leads to beatings, dismissal and even extermination. Does this sound like a society that understands stewardship?

Dr. John Kullberg

Ethical values indeed are not well served unless *stewardship, not ownership,* is the norm for companion animal care. In large part, because of an all too permeating domination attitude in our society, the need for shelters, spay/neuter clinics and, yes, sodium pentobarbital solution, continues to escalate, while free enterprise thrives upon ever-increasing market demands: the puppy mill industry, pet foods and such products as the controversial flea and tick repellent, *Blockade.* Shelters, spaying and neutering clinics, humane law enforcement and humane educational antidotes are not adequately present today to diminish the societal thirst for more and more pets and, consequently, more and more pet-related products.

Sharing our lives with companion animals must be seen as a *privilege,* not a right,[11] and a privilege that should be tightly and ethically regulated according to what's good for these animals, and not what a human argues might be good for humans. Given an overall commitment on the part of shelter workers and volunteers to more ethical standards for having and then caring for companion as well as other animals, some immediate remedial action can then be initiated, But we have to have that commitment first.

Shelters must participate in efforts to convince legislators to regulate companion animal breeding by individuals, including back-yard "miracle of life" breeders, professional breeders and puppy mill owners. Breeding must be severely discouraged, and spaying and neutering highly rewarded. A $50 dog license spay/neuter differential is not too high, in my view, and neither is a $100 breeding permit for individuals, and $1,000 or more for professionals, with permits available only when the applicant (you, me or others, if we're going to breed, which I hope we don't), can demonstrate that individuals and families are available for all the animals bred.[11a] And we must have heavy fines applied for noncompliance with dog license and breeding permit legislation. The day has come when we must recognize that pet breeding, except in exceptional circumstances, is to be seriously discouraged.

We must also discourage pet store dealers and other breeders who ply their trade in areas where shelters and pounds cannot find responsible homes for all the unwanted pets they receive each day, many of whom are perfectly healthy and of excellent disposition. These animals are killed because shelter workers and volunteers cannot find responsible homes for them, and because shelter cages and dog runs are constantly overflowing

## Responsibilities of Shelters

from the steady stream at the intake counter of more and more unwanted animals. Any of you here today, who have not spent ten minutes observing an intake counter at a shelter, should do it for the shocking experience it will give you. It will be probably the most ten disappointing minutes in your life. Estimates range as high as 15 million unwanted dogs and cats killed in the United States every year, and still, the dog and cat breeders breed and breed—and get richer and richer.

Shelter managers might well consider asking volunteers to leaflet those who trade at pet shops, letting these potential pet shop customers know that hundreds of perfectly healthy and well-dispositioned animals are waiting in nearby shelters for good, responsible homes. Pet stores work off supply and demand; the shelters do not. Shelters are forced to kill healthy animals. We don't have human baby shops. Let's stop having puppy and kitten shops.

Those who work in shelters must also fight the researchers, who look upon the tragedy of pet overpopulation and the consequent shelter and pound populations of unwanted animals as research laboratory resources.[12] On the federal legislative level, we need to become personally involved in lobbying for passage of New York State Congressman Mrazek's bill, the Pet Protection Act. But don't be lulled into a false sense of achievement even if this much-needed federal legislation is finally passed. The Mrazek bill will not replace the need for state legislation prohibiting pound and shelter animal seizure in those states currently without such legislation. The proposed federal law only applies to prohibiting the use of pound and shelter animals in research where the funding source is the U.S. government. State grass roots efforts to ban pound seizure are essential, with support from humane societies and through the use of resources developed and available through the National Coalition to Protect our Pets, if the disgrace of pound seizure is to be once and for all ended.

The pet industry grosses $7 billion a year. Add into this amount the pet *supply* industry (Hartz Mountain and others), and this figure approaches $10 billion annually. A 5 percent wholesale tax to fund the programs attacking the causes of pet overpopulation could bring hundreds of millions of dollars every year to community animal shelter and spay-neuter clinic needs, humane law enforcement efforts and humane education, including media-based tutorials that focus not on kindness—as good as that is—but on responsibility, stewardship, obligation.

Dr. John Kullberg

Those who operate or who would like to operate shelters and spay/neuter clinics should be leading the fight for this desperately needed tax. The reason free (forget about low cost) spay/neuter clinics are not all over Boston, New York and Chicago, for example, is that the politicians claim that they have no money to fund them. Let's help politicians find the money, and I submit that the quickest route to the kinds of money needed is a tax that would raise the price of a can of cat food by less than two cents. The pet food producers, who through their marketing efforts are in part responsible for companion animal overpopulation and related problems, should pay a generous share of the costs for long overdue and increasingly needed solutions. They will not do so in anywhere near an adequate way regularly and voluntarily. They should therefore be forced to help us through a dedicated tax.

Finally, shelters must work to counter some of the negative side of the human-companion animal bond studies. Reducing animals to vitamin pills is very dangerous.[13] If human-companion animal bond studies promote responsible care of animals, even if only because pet benefits to human health make pets worth more, fine. But if the studies promote the notion that everyone should have a companion animal, whether or not he or she can care for a pet responsibly, then we have even more serious problems ahead of us. It's no coincidence that the pet food industry glommed onto the human-companion animal studies as quickly as it could, and we should read something into that. Valuing animals on the basis of what they can do for us is the same as our valuing each other on the basis of what we can do for each other. That standard violates inherent worth and is wrong. We deserve more, and so do companion animals. Those who operate shelters must dedicate themselves to educational and legislative tasks that are aimed at the causes as well as the results of pet overpopulation, if we are to make significant and long-lasting gains in overcoming society's pernicious, inhumane attitudes toward animals generally, and pets particularly.

**VALUSEK:** Thank you. I'm sure Dr. Kullberg has given all of us a great deal to think about.

## Responsibilities of Shelters

**VALUSEK:** It's now my pleasure to introduce Kathleen Young, manager of Bide-A-Wee Home Association's Westhampton Shelter.

**KATHLEEN YOUNG:** Good morning. I thought I would start by explaining a little about Bide-A-Wee, and how and why Bide-A-Wee functions as a non-euthanasia animal welfare organization.

Bide-A-Wee Home Association was founded in 1903 in New York City by a Mrs. Kibbe. There's still a Bide-A-Wee shelter in Manhattan. It's just a few blocks from here right now, in temporary shelters on 53rd Street, and there are two other locations on Long Island as well. One's been in Wantagh in Nassau County, since 1915, and the other was opened in 1956 on the east end of the island in Westhampton. The association was chartered as a shelter for animals being given up by their owners, as an alternative to abandoning them in the streets of New York City. The charter also mandates that no Bide-A-Wee animal will be destroyed unless it's incurably ill, a fact that tends to make people much more willing to turn their pet over to a shelter.

The name, Bide-A-Wee means, in Scottish: "Stay awhile." Mrs. Kibbe took the name from the roadside inns she encountered while traveling in Scotland, which were called Bide-A-Wees. The idea was to provide a place for unwanted animals to stay for as long it takes to find homes for them. As with any organization, the fundamental principles on which Bide-A-Wee was founded necessarily put limitations on how the association can function. We could not, for example, do animal control, because of the need for euthanasia. Bide-A-Wee's charter also limits the amount of political involvement the association can engage in.

But, within the guidelines of its charter, Bide-A-Wee does provide a broad spectrum of services: In addition to the three shelters, each location has a veterinary clinic open to the public. The two Long Island locations have pet memorial parks or a pet cemetery.

Bide-A-Wee also offers various outreach programs such as humane education,[14] pet loss grief counseling, dog training, animal behavior counseling, pet-facilitated therapy community events, and more.

Kathleen Young

The fact that Bide-A-Wee has a non-euthanasia policy provides the public with a place to bring their animals with the confidence they won't be put down if they're not adopted. Unfortunately, there are people who abandon an animal rather than take it to a shelter that does put animals down.[15] Bide-A-Wee offers an alternative, and the animals get a second chance.

This would be an ideal situation if it weren't for the problem which is the theme of this conference: there are too many animals and there are not enough homes to go around. I can attest to the fact that at least 80 percent, probably closer to 90 percent, of the phone calls we receive at our shelter are from people that want to give up a pet. It's not too hard to figure out that if four or five of our customers want to bring in a pet, that leaves at the most one potential adopter for every four unwanted animals. Therefore, there is more often than not, a waiting list to bring an animal into Bide-A-Wee or any other non-euthanasia shelter. Animals are accepted by appointment so that the cage space will be available, and the veterinarian will be on hand to examine and vaccinate the animals before they're accepted for adoption.

While these limitations can cause inconvenience for the public and for the staff, they do give Bide-A-Wee an opportunity to be sure that the animals in our shelters are as healthy as possible. This is crucial in a non-euthanasia situation, because all the animals that do come down with treatable medical problems while they're in our shelters are treated until they're well enough to be adopted. From a practical standpoint, in terms of staff time, financial considerations, and just plain humane animal care, disease has to be kept to a minimum.

Animals with contagious diseases are not accepted. If a condition is treatable, the owners are offered medication free of charge and given another appointment for a date after the prescribed treatment period. Adult dogs are heart-worm tested and put on preventatives. All cats and kittens are tested for feline leukemia virus. All animals that are of age are spayed or neutered before adoption[4] and, of course, all animals are kept up to date on vaccinations, de-wormed and de-fleaed during their stay at Bide-A-Wee.

All these preventive measures help to protect the health of the animals at Bide-A-Wee and also to ensure that the public is offered the healthiest possible animals for adoption. Once an animal is adopted, we provide free health coverage at any Bide-A-Wee clinic for the first two weeks of ownership, in case the pet has been incubating a problem that shows up after it's adopted. We will also reimburse up to $50 of an emergency visit to another veterinarian if the Bide-A-Wee clinic is closed and the pet needs immediate care.

In addition to dealing with the basic fact that one animal must be adopted in order to make space to take one in, a non-euthanasia shelter also has the responsibility to consider the future of the animals it does take in. I'll talk about screening adopters in a few minutes, but before we get into screening, we must consider whether an animal even has a reasonable chance of being adopted.

Is it fair to take in an animal that for temperament, health or whatever reason, is likely to spend the rest of its life in the shelter? This could be unfair not only to the animal, but to all the animals that could use that cage or kennel run and be adopted from it over the course of one animal's ten- to fifteen-year life span.

That is not to say we don't take in our share of older pets and handicapped animals and behavior problems, some cruelty and abuse cases, and other hard-to-place animals. We do. And we also make exceptions for animals originally adopted from Bide-A-Wee. But it is important to remember that the more adoptable animals are the ones who will move out quickly and make room for the next ones to come in. So there has to be a balance struck somewhere, or the adoption operations could come to a standstill. With no animals adopted, none could come in.

This is one of the areas where a sensitive and knowledgeable person—be it a staff member or a volunteer—can be of great value to an organization and a service to the public. While Bide-A-Wee can't immediately help every person that wants to bring in an animal, we can and do offer advice and literature on how to deal with the problems.

Sometimes a behavior problem can be worked on and resolved so that the people can keep the pet. Sometimes they just need to be pointed toward a solution or given hope that there is one, such as dog training or veterinary help. Other times, people know in their hearts their pet isn't adoptable. Maybe it's old and it's incontinent, but they need to be reassured that it's okay for them to have that pet put down painlessly rather than spend the last part of its life on a concrete run in a loud, unfamiliar shelter full of strange animals. We have a responsibility to the animals to try to help these people in one way or another, and we do try.

Unfortunately, there are a lot of people out there who don't want to solve their problems. They just want to get rid of them. They either don't want to or they can't wait until we have room, and they won't put up with the animal any longer. With the unwanted pet population what it is today, a non-euthanasia shelter simply can't handle this problem alone. This is where cooperation and communication between non-euthanasia animal welfare groups and municipal animal control facilities can be so valuable.[16] We happen to have a very fine animal control center in Southhampton County, where the shelter I run is located.

It makes a tremendous difference to be able to assure a person that, while we can't give them the immediate help they want or need, their animal or the one they've found will be taken care of properly at animal control, that it will be given every possible chance to be adopted or claimed, and if worse comes to worst, the animal will be put down in a humane manner with kindness and a minimum of stress. I realize this isn't the case at all animal control facilities, but it proves it can be done. The fact also remains that if euthanasia shelters did not exist, there would be no place for surplus animals to go.

Until such a time, if it ever comes, that there are no "excess" pets, somebody is going to have to do the nasty job of putting them down. It only makes sense to try and communicate from both sides of the fence, so to speak, and see what we can all do together to improve the situation that these animals are in, through no fault of their own. Finger-pointing and accusing each other of various extremes from "killer" to "humaniac" isn't going to help the animals. Dialogue and cooperation in order to reach an understanding and work toward a solution is the only reasonable way.[16]

I can't very well talk about the responsibilities of shelters and not mention screening of adopters. At Bide-A-Wee, we have a written adoption application which asks about 20 questions on the person's lifestyle, what they want in a pet and what they can offer the pet. We also ask for personal references, their veterinarian's name, their landlord's name and number if they rent, identification at the address they put on the application and, of course, how they feel about spaying and neutering their pet.

The two biggest advantages to having a written application are that it prevents the problem of the person doing the screening forgetting to ask an important question, and it gives us the person's responses in their own handwriting, so they can't later say we misunderstood them—they didn't mean that. It also facilitates building up a list of people requesting certain types of animals so we can match them up in the future if we don't have what they're looking for at the moment.

There are certain conditions that invariably must be met before Bide-A-Wee will release an animal:

- We must be absolutely certain the animal will be spayed or neutered, and it's paid for in advance. If there is any doubt about whether it will be done, and the rest of the application is fine, then only a spayed or neutered pet is released.

- The person must have ID with their current address, no exceptions.

- A surprise gift for another person, except their own child, that lives with them, is not allowed.

- The person must be 21, have a responsible, reliable source of income, and have permission of their landlord.

- They have to be head of the household, or, if they're not, we must contact that person for their approval and consent.

- They must be willing to provide the proper food and shelter and veterinary care.

- Guidelines for how and where the pet is going to live are geared towards being sure that the animal is a member of the family, lives in the house and has ample opportunity to interact with the family. Young puppies and kittens shouldn't be living with toddlers or left alone all day, and no animal should be allowed to roam outside unsupervised.

Exceptions to some rules can be made by qualified personnel, which is why interpersonal skills and good judgment and training of the adoption staff are extremely important. For example, an unhousebroken St. Bernard isn't going to be expected to live in somebody's living room, but we would expect it to have a fenced yard,[17] a warm, dry shelter, shade for summer, proper food and water, vet care, and a chance to interact with people on a regular basis.

Just as we are as thorough as possible in screening out unsuitable homes, we need to be scrupulous and need to be sure not to screen out potential good homes without making an attempt to educate the people. Take, for example, the person who indicates they won't spay their dog until after she's had one litter. They may just not know that it's healthier for the pet to be spayed before her first heat. Maybe they think they're doing the right thing. Through a little dialogue we might find out they are actually relieved to find out they don't have to have puppies. Or, if we're not convinced, then we don't give them an unspayed animal[4]. It's basically a judgment call on whether the person's really listening to you and learning something, or just "yessing" you. We like to make these decisions as a committee whenever possible.

Most people who want a pet badly enough will get one somewhere, if we don't adopt to them. So, of course, we attempt to try to educate and communicate with them. I believe the most difficult part of the process is how to instill in a person the feeling that when they adopt a pet, they are accountable for that pet for the rest of its life. For that reason and others, the importance of follow-up on adoptions can't be overemphasized.

At Bide-A-Wee, we try to place a call to each adopter about one week after adoption to see how things are going. If there are problems, we keep in touch until they're worked out. Bide-A-Wee provides dog training and animal behavior counseling. Training crates can be borrowed and used under professional supervision. And we offer just the moral support of being there to answer questions when they come up. When spay/neuter is due, reminders are sent and follow-up calls are made when needed to make sure it gets done.

Even with the best screening policies, some animals are going to be returned to the shelter. While it's not the best thing for the animals, if we have kept in touch through follow-ups, it being returned might be the worst thing that happens to it.

Our follow-up records are also kept and referred to when a person comes in to adopt a pet. If they're previous adopters and have brought us pets before or were previously refused a pet, this information is available to us, and we can act accordingly.

## Responsibilities of Shelters

While a non-euthanasia shelter by itself can't solve all the animal problems in the community, I do think a responsible non–euthanasia program can provide the best possible environment for matching up adopters and pets. We know as much as we can about the animal, its health, its temperament, it's history. And about the people through screening and education and follow-up.

Sounds like the ideal place to work, right? People who work in non-euthanasia settings do have the opportunity and the freedom to be very careful and to ensure to the best of their ability that every animal will have a chance at a healthy life in a responsible home. They also don't have to deal with euthanasia on a regular basis. But people still burn out in this business. The frustration level is so high, and the public ignorance about responsible pet care is just so widespread, that people feel they are not making any difference. For each animal placed, there are so many more waiting to come in.

All I can really say, and it's been said before, is that there aren't any magic answers. We can just try each day to do what we can to educate somebody, to communicate a little and maybe create one more responsible home. And we can support each other. There's too much to do to waste time fighting among ourselves. Maybe as individuals and as organizations, we can only take tiny steps at a time toward a solution for the entire animal problem, but at least we can do our very best to make sure we're not contributing to the problem.

**VALUSEK:** Thank you. Are there any questions?

**VOICE:** I want to make some comments concerning what John Kullberg was talking about. What he's saying is true. Why don't we reclassify the station in life of the dog and the cat? In other words, let's reclassify it law-wise. For instance, if a dog is outside and he doesn't have any water or anything like that, he has as much right as if the guy had his little boy out there. Let's reclassify the dog's station in life. Instead of being a property, fourth class citizen, let's make him a second class citizen.

**KULLBERG:** Well, I think what you're talking about is what I mentioned—the difference between ownership and stewardship.[11] In the United States, dogs or cats are property. In the United States, the law now recognizes dogs and cats as property, as objects to be owned.

I might add how this problem is causing some grief right now. A person who wrote from The Fund for Animals asked us to take the pets of the homeless and find good responsible homes for them. I don't think this would be legal. Those pets are the property of—and are therefore owned by—homeless, and for me to take the pet and to place it with someone who has a home would put us at legal risk. While I think we must move away from ownership to stewardship, this means reclassification in the law of how animals are understood, and also raises the question of who would make what decisions regarding stewardship, using what criteria.

**VOICE:** That would require legislation.

**KULLBERG:** It absolutely would require legislation, in that the animal today, the pet today, is recognized as a material object by the courts and by the State. I know many legislators well, and a lot of them play very cute. I think they want to do what they can to alleviate animal problems, but they also, ultimately, are politicians, and they choose what they publicly support very carefully. That suggestion I have on the pet food tax, which some of you might like and some of you might not, when I spell out the details, usually most legislators will *privately* support it. But it's a tax, and legislators don't like dealing with a tax, and their *public* support just isn't there.

In the next 20 years, though, I anticipate that value-added taxes will increasingly be the normal route to raise money. The federal income tax has gone far enough. The state income tax has gone far enough. But we have to raise significantly more money from the public for good causes. We're leaving a horrible, horrible inheritance to our children, an inheritance called the national deficit! We are going to have to raise more money through excise "nuisance" taxes. It's the way it's done in Europe. Every time you go to a theater, every time you buy toothpaste, every time you rent a car, whatever, there is a dedicated tax. In our lifetimes, I predict we will see this. Senator Biden right now is running for president on a platform that promotes more and more of this concept in the United States, and I just want to get in there first for money to help needy animals. I want the dedicated tax on pet food to get us the kind of money that we need for animal care and control programs in this country, as soon as possible.

**VALUSEK:** Are there any further questions?

**VOICE:** What do you think of shelter personnel breeding dogs?

**KULLBERG:** To me, that's wrong. One of the real problems we have—I have, you have, and others have—is not enough introspection as to where our own values are.

We all have attitudes that we haven't looked at. One might be, "Newfoundland is a great breed, and I'm going to breed them." Another might be, "Everybody should have a dog." Another might be, "A dog is better in a home than euthanized, regardless of the home." There are a lot of different attitudes, and there are a lot of different problems as a consequence, so I think that the move to clearing the air in the humane movement begins with ourselves. And that's why the first portion of my comments today said if we get our own act together our own philosophies straight and our own ethics properly in line, only then can we be effective in the legislative and educational realms. And I think all of us—I, you, and everyone else—have a long way to go.

**SHERI TRAINER:** I'm from the Bide-A-Wee Home Association. I'm a little confused and concerned about the direction that animal control is going to be taking in New York City, and I have not been keeping up with the latest legislation. I would like you to clarify the new legislation that was passed concerning bidding for animal control amongst the five boroughs.

**KULLBERG:** Thank you for asking that question. We fought like the dickens to get this legislation passed, for one reason only. Every time I went down to try to negotiate a contract in the City of New York in the Mayor's office, they would all but

## Dr. John Kullberg

laugh at us. Rightly so. This crazy law was in place that somebody thought was great back in the late 1800s, that in essence said if animal control is going to be done in New York City, the ASPCA had to do it. We think the law says we'll do it, if, in fact, as a not-for-profit, we choose to do it. But we were told by our legal counsel that it would take years to resolve the difference between our legal interpretation and the City's. As a consequence, City Hall pretty much has been telling us what they would give us, and I couldn't come back and say: "No, you give us what our programs really cost, or we're going to close our doors." Legally, we were in a bind. In service-related negotiations, you have to be able to fall back on a stoppage of service threat, whether you really want to close your doors or not. What the recent revised dog license law did was simply clarify things in our favor. We consider it a pro-ASPCA, pro-animal law.

The law now says that with 18 months' notice, I can tell the City, "We're not going to do it anymore." *Quid pro quo*, the City can tell us with 18 months' notice, "You're not going to do it anymore." Now, the City can contract with us or with someone else, and I think that's fine. If I were the mayor, I would want to be able to decide where the monies are going to go.

If Bide-A-Wee, for example, wanted to get into animal control, what's wrong with that? But I don't think this is going to happen too quickly, because you see, I just did some calculations: We are in effect writing a check to the City of New York this year for over $1.5 million. That is our donation to the City for the privilege, essentially, of killing its excess animals, because that's basically what we have to come up with to keep our shelters open and our ambulances on the road. Although our adoption rate is 20 percent, the remaining animals are killed because there aren't any homes for them. Now, we are willing to perform animal control work for the long term, but you heard about Los Angeles. The government there puts up shelters, spay/neuter clinics and then pays the full cost of operating them. We, too, should be getting paid for it. We should get full operating costs. But New York City has never even given a penny toward the construction of shelters for the City's unwanted animals, and even after we build the shelters, the City doesn't give us enough money to operate them!

We have just announced plans to sell all of the ASPCA's property in Manhattan. We are going to spend more than $20 million in reorchestrating our entire program in Manhattan. We're committing in excess of $5 million dollars for a new animal shelter in Manhattan. So, obviously, we are here for the long haul.[6] But if another humane society, either in Manhattan or in a borough where there is no shelter, would wish to make application for an animal control contract with the City, we'll help them out. We don't want a monopoly. We think, therefore, it's been a very good legislative change that now allows other organizations to consider animal control programs in New York City.

But don't misread the legislation. Although we believe that ten years, twenty years and fifty years down the road, you will find that we're still intimately involved, as Gretchen Wyler said, you should get rid of the notion that the only people who can do animal control and do it humanely are people associated with humane societies. Nonsense! Take a visit to Washington, where in fact it's done by a humane society, but the City pays the full price. Visit Chicago, where they have a Taj Mahal of a shelter. Look at what's going on in Los Angeles. The cities can run animal control programs superbly, in part because humane societies such as ours will continue to keep our eyes on them. Gretchen, do you think Bob Rush, the manager of the city shelter system in Los Angeles, doesn't know that humane societies are looking over his shoulder? This does a lot to ensure humane city-run shelters.

**ANN FREE:** For clarification, are you referring to the excise tax as a national thing? Would this be federal legislation? And the other part of the question is, would this not be similar to the excise tax on arms and ammunition?

**KULLBERG:** Well, you have value added taxes today on arms and ammunition, gasoline, tobacco, and liquor. A value-added tax on pet food could be done beautifully at the federal level, and the funds collected apportioned to the states on some formula which could be based on the total of the dogs and cats registered around the country. What proportion a given city's pet population is of the national total would be the share of the monies the city would get from the total monies collected, minus expenses. There are a number of other fair ways for apportioning these tax revenues.

I don't see a pet food tax at the national level first; however, I see it at the state level. I would like to see such a pet food tax bill passed in New York State as a model bill which other states could then evaluate. And I might tell you that Bob Rush, Dr. Palmer (the official who runs the animal control in Chicago), and the person who runs the animal control in Austin, Texas—the four of us have an informal bet as to which one of us will get such a bill passed first, because we have been pushing for such a bill in our local areas for a long, long time. I'm hopeful that it will first be passed in New York, but I don't really care where it is first passed, as long as it gets passed. We'll then have a model state bill. The monies will start coming in, and the better programs for animals can thereafter take place.

**FREE:** Who's fighting it? How about Kal Kan and the pet food industry?

**KULLBERG:** Oh, yes, I hear from them every once in awhile, and they certainly will pull out the lobbyists in grand style to defeat any proposed pet food tax bill.

**FREE:** People don't have enough garbage, so the dog food companies don't have any competition, do they, for feeding animals?

**KULLBERG:** And they have a lot of money. But I think that the logic of something can dominate over time and, after all, we're talking about your demand for more animal care and control programs, for better humane law enforcement and education. If your demand is great enough and the legislators have to provide it but they don't have the money, they're going to look for a money source. And all the lobbying Kal Kan and others undertake won't be effective.

# What's Wrong with Pound Seizure?

**SAMANTHA MULLEN:** Our next speaker is John McArdle, an anatomist. Dr. McArdle is Scientific Director of the New England Anti-Vivisection Society in Boston.

**DR. JOHN McARDLE:** When I was first approached about discussing this topic, New York State was still practicing pound seizure in certain locations. Pets were still being released to laboratories. Only a few, but it was still a possibility in New York State. Since then, I was very pleased to learn that new legislation passed in the state,[8] makes New York the twelfth one to eliminate the practice of releasing pets from shelters for research. But it is far too soon for New Yorkers, residents of this state, to become complacent about pound seizure. As John Kullberg said this morning, it is, in fact, still a problem in the State of New York. It's a problem because the legislature did not ban importation. You have only a few places in this state that release animals. And places that were doing it were guaranteeing 100 percent of the animals were owner-surrendered pets. That was rather unusual. Most animals that have been used, particularly here in the New York City area come from New Jersey and Pennsylvania. So until you've actually stopped importation, you haven't eliminated the use of pets; you've simply transferred the problem to another geographic location.

As John mentioned earlier, there is national legislation to deal with this issue, both in the House, and in the Senate. I think it's very important, since you now live in a state that has banned pound seizure, that you let your representatives in Congress know that the state is on record now, as being opposed to it. I think it's important that those representatives co-sign the legislation, and work for its passage nationally. It *will* make a major dent. Particularly the Senate bill may very well finally put the Class B dealers out of business.

## Dr. John McArdle

I want to talk a bit about the history of the use of pets in research. Where does it come from? How far back does it go? And about the pivotal role that New York State has played, here in the United States, in the use of pets in research.

Take a look at vivisection 150 years ago. At that time, in the early 19th century in Europe, the use of animals in research was a relatively minor enterprise. There weren't many people involved in it, and it was limited to certain universities. But the term "vivisection" was very real. Anesthesia did not exist. You had individuals, such as Magendie, in Paris, and later his student, Claude Bernard. In one series of experiments by Magendie, over a thousand puppies were used. These animals were either tied or nailed to a piece of wood and then they were operated on without anesthetic—major surgical operations. It was quite common to have dogs and cats used in research. When they wanted them, they didn't have to worry about going to a shelter; they simply went out and picked them up off the streets.

Anesthesia became available in one form or another in the late 1840s, and early 1850s, for humans. It was much later before it was used with animals. So very early on we have a conflict between those who are using animals for research, the animal welfare organizations, and anti-vivisection groups. In the initial phases, the cruelty, the suffering, of the animals in the laboratory was *very* real.

In the late 1800s, at least that's the way I read the history, you started, initially, getting some benefits from animal research. Things were coming out of the laboratory that were of practical human benefit—things that weren't just for curiosity's sake, knowledge for knowledge's sake. An informal truce was formed at that time between those who were opposed to using animals in research and those who were conducting the research.

I believe at this point we had a clear distinction within the animal welfare community between those who were concerned about the care of animals and those who were concerned about their use. The fact that animals were used at all in research continued to be the concern of the anti-vivisection societies. Many of the animal welfare organizations were concerned about care, or became more involved in companion animal issues.

In England in particular, we have passage of the 1876 Act that dealt with laboratory animals. It had very specific provision for the protection of pets but did not include horses.

Now, an interesting incident occurred. Although the Act prevented multiple surgeries on animals, the use of animals for gaining manual skills, a fairly well documented case was discovered, in which a small dog was used. And it was used for repeated procedures. So when this was reported, among the many things that happened, one of the neighborhoods in London, Battersea Park, made a small statue of a little brown dog, and put it in the park.

The response of the medical community at that time (we're talking about the turn of the century) was first to go to the Council and demand that it be removed. (It was interesting that there was an alliance at this point between animal welfare groups, anti-vivisection societies, working class people and Suffragettes. They were all supporting this statue.) When going to the Council didn't work, medical students from University College went across the river, and rioted with people of the community to try to have the statue destroyed. It was later removed. It's only in the last few years that a new one has been created and put back into the park.

Now, if we look at what's going on in the early 1900s in the United States with the use of pets, again, research is still a relatively minor enterprise—a few individuals, primarily at the universities. Individuals such as Banting and Best in Toronto, beginning their initial work on diabetes and insulin—who have admitted openly that when they needed a dog, they simply went out and took it off the streets of Toronto.

There was a gradual increase in the first half of the 1900s, in the United States, in the amount of biomedical research that was going on and in the number of individuals that were involved in it. The first law specifically dealing with the use of pets in research was passed in Pennsylvania in 1917, prohibiting the release of dogs from shelters. I think it's interesting that it was that early that there was an objection to this practice.

Events have changed drastically since the end of World War II. If you look at what was happening from 1945 to 1950, when the National Institutes of Health, and National Science Foundation were established, there was a large amount of government money becoming available to support funding for research. There were large numbers of individuals going through college in the late '40s, many of them GIs coming back from the war, getting Ph.D.s, D.V.M.s, M.D.s. So there were two parts of the triad: money to support research, and individuals who were entering careers in research.

What did not exist in the late 1940s and the early 1950s were commercial breeders of laboratory animals. Charles River, the first and now the largest, began in 1947 supplying local laboratories. Beginning in 1947, the medical community looked around and said, "Where can we get a guaranteed supply of cheap animals?" The conclusion was that: "The local shelter always has too many." That's really about the extent of the analysis that was done on the situation. I believe that by declaring war on the animal shelters, the medical community broke the truce that had existed between humane societies and biomedical research.

We, those of us who have been involved in shelters, have always viewed our facilities as a sanctuary—I certainly hope you have. You've never viewed them as a warehouse, or someplace that supplies subsidized products for biomedical research. The research community, very early on, is on record as viewing this as a renewable resource, a very large and extensive renewable resource. They have never really shown any serious concern for humane issues, as far as the things that we are concerned about: pet overpopulation, spaying and neutering. To the research community, particularly in the late 1940s and early 1950s, this was a supply and demand situation. They needed animals for research. They couldn't breed enough of their own, commercial breeders weren't there, and people who were running the shelters had more than enough.

So they began passing laws around the country, at the state, county, and city level. There were polls taken in the newspapers, and people were very supportive of the use of shelter animals in research at that period in time. Of course people didn't know much about what was going to happen to the animals, and they certainly didn't know about the negative consequences. There was no organized opposition to the practice. The National Society for Medical Research, an organization that fortunately no longer exists, was formed at the time, primarily to promote pound seizure laws. That was one of their major purposes. New York State was one of the first states to pass a law requiring pound seizure.[18]

Now, what are the problems with this? Pound animals, both

## What's Wrong with Pound Seizure?

cats and dogs, were used in very large numbers in the early 1950s. What's happened over the last 37 years is that they've gone from being a "major resource" used in large numbers, to now being less than half of one percent of all the animals used in research laboratories. They're a relatively insignificant resource at this point and I'll emphasize in a moment why I think that's important. But one thing that they never did in the early 1950s was conduct a scientific study to determine whether these animals were suitable for use in research or that they should, in fact, be used. They just simply said: "They're there and we can get them."

What's happening today is basically a continuation of an old tradition, a perception of cheap animals which inhibits the development of more suitable alternatives. There is no major area of biomedical research at this time that is, in fact dependent on the use of pound animals. Now, I don't usually read long quotes, but I have one here that I think is particularly important.

One of the most eloquent statements that I've seen on this issue of the relationship between medical research, pets and shelters was made in 1952, by Robert Gesell, who was professor and chairman of the department of physiology at the University of Michigan. He made the statement at a meeting of the American Physiological Society. It says:

> *The National Society for Medical Research would have us believe that there is an important issue in vivisection versus anti-vivisection. To the physiologist, there can be no issue on vivisection per se. The real and urgent issue is humanity versus inhumanity in the use of experimental animals. But the National Society for Medical Research attaches a stigma of anti-vivisection to any semblance of humanity. Anti-vivisection is their indispensable bogy which must be kept before the public at any cost. It is their only avenue towards unlimited procurement of animals for unlimited and uncontrolled experimentation. The National Society for Medical Research has but one idea since its organization, namely, to provide an inexhaustible number of animals to an ever-growing crowd of career scientists with but little biological background and scant interest in the future of man.*

Dr. John McArdle

> *Consider what we are doing in the name of science, and the issue will be clear. We are drowning and suffocating unanesthetized animals in the name of science. We are determining the amount of abuse that life will endure in the name of science. We are producing frustration ulcers in experimental animals in the name of science. We are observing animals for weeks, months, or even years under infamous conditions, in the name of science. It is the National Society for Medical Research and its New York satellite that are providing the means to these animals. And how is it being accomplished? By undermining one of the finest organizations in our country, the American humane societies.*
>
> *With the aid of a halo supplied by the faith of the American people in medical science, the National Society for Medical Research converts sanctuaries of mercy into animal pounds at the beck and call of experimental laboratories, regardless of how the animals are to be used. What a travesty of humanity! This may well prove to be the blackest spot in the history of medical science.*

Many of you may not have heard of Dr. Gesell, although I suspect you all know his daughter, Christine Stevens.

What is the current situation on pound seizure? It's now banned in 12 states.[18] This year, there was activity in more than 20 states that I am aware of, and I think it'll be even more next year.[18] I believe it's important to point out that the current levels of anti-pound seizure activity, whether at the state or local or the federal level, began with the efforts here in New York to repeal the Metcalf-Hatch Act.[19] You people got it going a number of years ago, and it's been snowballing ever since. The use of these animals is banned throughout western Europe, as of earlier this year. The European Common Market countries are not to use these animals in medical research. The International Organization of Medical Sciences, which is an affiliate of the UN, has come out against the use of these animals, as of late last year. The World Health Organization says, "Don't use them." The National Institutes of Health, which is the largest biomedical research entity in the world, has an in-house *de facto* ban on the use of animals from shelters. They use a few random source animals, but they make a point that none of them come from shelters.

## What's Wrong with Pound Seizure?

So we see increasing activity at the state and local level. (Even the passage of the federal bans, which would prohibit their use in Federally supported research, would not stop their use in research not Federally supported, or in teaching.) We're going to have to stop the use of these animals. They'll still want to use some of them for teaching. So I think it's still important that you get involved.

What are the effects of pound seizure on shelters?[18] Hopefully, you're all aware of this, but I think it's important to review some of the information on how pound seizure directly affects your ability to do what you do best, take care of the community's pets and educate the community's pet owners.

There has been a letter available for many years from Bob Rush in Los Angeles, making it quite clear that one of the major reasons they banned pound seizure in Los Angeles was to restore community confidence in their shelters and to eliminate an activity that was dividing, putting a wedge between the public and the people who are there to serve the pets of the community and pet owners.

We know how pet owners feel, from informal surveys in Salt Lake City, where they've been fighting pound seizure for a number of years, and where I think we're going to win very early next year. One shelter surveyed owners as they were going out, as to how they felt about it. Universally, they were opposed to it. In fact, they weren't going to cooperate if the shelter stopped its policy of not releasing animals for experimentation. That particular shelter, in violation of its own state law, has never released animals, although the university comes in weekly and demands them.

One of the most interesting surveys that's ever been done, and it's one that could easily be repeated in other parts of the country—I hope it will be someday—was done in Cincinnati a number of years ago. In Cincinnati, the Hamilton County SPCA is a rather unusual institution, because the Hamilton County dog warden is also the Hamilton County SPCA. It's the same building, it's the same trucks, it's the same employees wearing the SPCA uniforms. So if an individual in the community in Cincinnati calls about a stray dog that they've found in their back yard or they've picked up, a person comes out in an SPCA truck wearing an SPCA uniform, and takes it back. Because it's a stray, it goes into the dog warden's side of the building, not the humane society side of the building.

## Dr. John McArdle

They keep all these records. Since Ohio is a mandatory pound seizure state, some of the dogs that go into the dog warden's side end up in research laboratories. The ones that go into the Humane Society side of the building do not.

Local activists went in and checked the statistics. Where someone had called on a stray, and the stray had been released for research, they went back to those people in the community and polled them, asking: "If you had known that by calling on that stray, the animal was going to end up in a research laboratory, would you have called?" Now, the sample size is not large, since they were stopped before they had a chance to finish the survey. At that point, however, 100 percent of the responses were: "No, we would not have called about that stray. We would have let it go, or would have tried to find a home for it ourselves." Even if people still come in to adopt animals from a shelter that releases animals for research, what are communities losing in animal control because people won't call on a stray?

Taking care of that problem here in New York will not take care of it nationally. People's pets are still being used in research institutions in this state. I don't have the time to go through a detailed document that I have here from the Michigan Humane Society, comparing the costs and benefits between shelters that release and shelters that don't. I'd be happy to send you a copy of that if you'd like one. It's a comprehensive explanation of how you can, in fact, increase revenues by stopping pound seizure.

Pet theft is still a real problem. Pets are still being stolen and sold to research laboratories, [20] although that is not the only reason they're being taken. But until we're able to eliminate the use of the pet-type animal from research and teaching laboratories in this country, we'll never be able to eliminate the threat of stolen animals. There is no need for these animals for teaching. There is an alternative available for every single one of the current uses. I make that as a flat statement because I have yet to encounter one where there was not already an alternative. And in places such as England, it is still illegal to use animals for that type of teaching.

What is happening with the research community? They have maintained their tradition of 37 years of being basically uninterested in our problems.[12] They're using the shelters as resources. You don't find them offering to help with education. They are not out using their educational resources to

## What's Wrong with Pound Seizure?

help us with the community, working with pet owners. You don't see them opening their clinics on weekends to do free spay/neuter operations for the communities. I received a letter about a week ago pointing out to me: "Please don't say that shelter animals are free when the universities take them. They are not. They cost us a great deal of money: to pick them up, to care for them." Most of the pound seizure laws that were passed were quite specific. They're going to get the animals for $3, $4, $5. In Ohio, it's still $3 for a dog or cat, and that was part of the original law, so they're guaranteed to get them cheap. I don't know how many of you could pick up, maintain an animal for a holding period and then transport it for $3 an animal. I suspect you'd have trouble doing that.

The humane community has been subsidizing biomedical research use of animals for decades. At the very least, you'd think that the research community would want to pay their fair share of community animal control costs, even if they're not going to help us with the pet owners.

What about claims of public support for pound seizure? You're seeing more and more of this. Particularly now that the national bills have been introduced, we have seen quite a bit of publicity claiming that the public supports the use of animals in biomedical research. The first of these claims, and the one they're still using, which caused trouble in Los Angeles County, was the claim that in Jackson County, Michigan, 83 percent of the people who brought an animal in, supported releasing that animal for medical research. I didn't buy that, because the exact same month that came out, another study came out saying that two-thirds of the people in the United States didn't even know animals were being used in research. So someone went in to take a look at those statistics on the owner option forms. They had six different options on it. And what they found was that, although there were over a thousand counts in the medical survey, there were only 600 people who actually brought animals in. If they brought in six, they were counted six times. They left out several hundred pet owners who came in, and forgot to sign the forms.

What's interesting about Jackson County, Michigan, is that it's an economically depressed area. If you bring a pet in, if you check off euthanasia as an option, they charge you for it. If you check off research as an option, the animal goes in free, there's no charge. So people bringing in a stray, people bringing in a litter, weren't going to check off euthanasia. What was actually found was large numbers of people checked every single column. Out of the entire sample of 1,059, there was one individual who checked the single column: "send the animal to research." That's something less than 83 percent. It's more like 0.5 of 1 percent actually supported it.

Gwinnett County, Georgia, used this statistic, claiming 80 percent of the people checked the forms in support of pound seizure. There never was a form in Gwinnett County, Georgia, and Gwinnett County, Georgia, banned pound seizure last year. In San Bernardino, California, there was a referendum last year. It was close. I'm happy to say that very soon, in fact it may have already happened, pound seizure is going to be banned in San Bernardino. Because the people there don't really support it, it's being stopped. Other surveys have been done, in places like Charlottesville, Memphis, Toronto, areas of Michigan, all of which show that the public is clearly opposed to the use of pets in research.

The National Association for Biomedical Research had their own national poll commissioned, and they selectively took certain answers out of it and gave those to the press. I obtained a copy of the entire report. There is a section where the public seems to support the use of strays in research. But there is another question that was asked about the use of pets in research, and 70 percent of the interviewed people said: "Under no circumstances should a pet be used in research." This distinction is important. The medical community would love to have people think that they're taking strays, in the sense of feral dogs.

You've all had experience with feral dogs and feral cats. They're going to cause you problems. I almost had my hand shredded once by a feral cat, so I know what they're like. The researchers won't take them either. They're not going to use them. They're going to take the pets. They're going to take the ones that are well-adjusted, that are in good health. They're easy to handle, short-haired, in a certain size range. If an animal comes out of a shelter to a laboratory, it was somebody's pet. That's important to keep reminding people. This *is* a pet issue, it's not a research animal issue.

What about other options? The research community, I think, is seeing the handwriting on the wall, and they're beginning to promote this as an alternative: Several universities are starting to talk about opening their own shelters, so people can come and bring the animals in. Rather than surrendering them to the local humane society, local animal control, they can surrender them to the shelter at the university. I'm not quite sure how the federal law is going to affect that. I think a little bit of publicity may stop it before it gets started. The University of Wisconsin in Madison, in particular, has been talking about that.

What other options? Well, first of all, if you pick up a stray, nobody knows who the owner is. Does that mean they don't use strays? What about people who bring an animal in and feel guilty about it? You've talked about people who are irresponsible who bring in these animals and who don't care about them, but what about the people who move into a home where they're not allowed to have pets? People who have to move into an apartment complex where they're not allowed to have pets? They may bring an animal in and feel very guilty about it. One way to expiate that guilt is to see it go for a good cause, for medical research. Part of the problem with that is that the average person is still unaware of what will happen to these dogs.

There really is no informed consent. If a member of the general public comes in and you say: "We're going to take your dog, and put it an experiment. We're going to burn it over 80 percent of its body and keep it alive for two months and then kill it," that's a lot different than saying: "We're going to do a study and then give it a good home." So unless you can tell them exactly what's going to happen to their pets, there's no way they're going to be able to give informed consent.

But why are they continuing to fight us? What's the real issue here? The issue is no longer pound animals, shelter animals, It's just not that important to them. The real issue is now the domino theory. This is the camel's nose in the tent. They're convinced that we're going to shut them down completely. There's also a problem of tradition. Some of these people have always used dogs and cats, and are not willing to learn a different way. Further, there's an inability to admit that, 37 years ago, they did something wrong and never should have started using these animals in the first place. Lastly, and this is one I'm encountering more and more, is simple medical arrogance.

Now, it isn't going to endanger research if we stop this. Let's look at government statistics. If you look at the states in terms of which ones have the largest number of research facilities, three of the top four are in states that ban pound seizure. If you look at the states that have the largest numbers of animals used in research, the top three states all ban pound seizure. If you take it another way, looking at the states that use the largest number of dogs in research, three of the top four are states that ban pound seizure.[18] In each of those cases, in the state that doesn't ban it, people are actively trying to stop it.

I want to leave you with this concept: Let's assume in 1950 that something very rational was done and a law was passed banning the use of pet animals across the nation. There's no way that the medical community can convince me or you or anyone else in this country that they would not have found some way to do medical research, that they would have not found some way to get the animals they claim to need for medical research, and that they would not have found some way to get the money, if we had stopped this 37 years ago. With the bills in Congress we have a chance to stop it now.

**VALUSEK:** We can take a couple of questions.

**PAMELA MARSEN:** I am humane educator for the Bergen County Animal Shelter Society of New Jersey, and my question applies to New Jersey. You said many of the New York laboratories are acquiring animals from New Jersey and Pennsylvania.

**McARDLE:** From dealers.

**MARSEN:** From dealers?

**McARDLE:** The largest dealer on the east coast is in Pennsylvania.

**MARSEN:** But in New Jersey, pound seizure is banned.

**McARDLE:** I know, but these dealers sell amongst themselves. You see, you still have dealers in New Jersey supplying research laboratories in New York City.

**MARSEN:** And where do they get them?

**McARDLE:** Mainly from dealers or auctions in states that don't ban pound seizure.

**MARSEN:** Where do the dealers get them, from auctions and—?

**McARDLE:** They go to auctions.[20]

**MARSEN:** In other states?

**McARDLE:** You can buy pet animals at auction in Pennsylvania.

**MARSEN:** But are any animals originating in New Jersey? That's what I want to know.

**McARDLE:** You mean from shelters?

**MARSEN:** From anywhere.

**McARDLE:** I wouldn't be a bit surprised if people actually sell them. They bring them in, they may be stealing them, that's possible. But not from shelters, legally.

**MARSEN:** I spoke to a young lady yesterday here. She's a reporter for *Glamour* magazine. She's doing a column for them. She told me her sister is a veterinary student at the University of Pennsylvania and her sister feels bad about animal experimentation, but said most of the animal diseases and the injuries that we have treatments for, we have because we have

Dr. John McArdle

done experiments on animals, and I'd just like to know how to respond to that.

**McARDLE:** That's a little broader issue than just pound seizure. It's the first defense. DeBakey made the same comments, and they say that every major medical advance in this century—sometimes they go back to the last two centuries—has come from animal research. That makes a mockery of almost all the material that's come from clinical research, which has been a major contribution. But if we take a simple measure of the Nobel prizes in physiology and medicine—now presumably there is some correlation between good quality science and whether or not you get a Nobel prize; I mean, we always assume that—going back to 1901 right up to 1987, two-thirds of all the Nobel prizes went to people who predominantly or entirely used alternatives to laboratory animals, right from the beginning. So I don't think you can make the statement that every major medical advance has come from animal research.

**MULLEN:** Thank you very much, Dr. McArdle.

# What Part Can Veterinarians Play?

**MULLEN:** The next topic is entitled "What Part Can Veterinarians Play?" It will be addressed by Dr. Arthur Baeder, of Rahway, New Jersey, and by Dr. Robert Case, of Schenectady, New York.

Dr. Baeder was asked to come to the conference to speak to you about a specific animal population control program in the state of New Jersey. He was instrumental in getting the state legislation passed which led to the implementation of that program.

Dr. Case is Legislative Chairman of the New York State Veterinary Medical Society. We've worked together on a number of issues. Dr. Case will discuss various approaches that can be taken to address the problem of controlling surplus animals. The approaches he will outline are less specific than those in Dr. Baeder's topic, which is geared to a particular program that has been recently begun in New Jersey. Dr. Case's remarks will be applicable to communities all over the country.

**DR. ARTHUR BAEDER:** I want to thank you for inviting us here from New Jersey to discuss our program. Sitting here listening to everybody talk, there are lots of problems, lots of causes of overpopulation. From the solutions that have been presented, legislation seems to be very, very important, and education seems to be important. In New Jersey we recognize this. We felt that there was one other thing that was really the key to the whole thing, and that is co-operation—co-operation among all the groups that are involved with animals, and especially, in this case, animal welfare. So, with this in mind, we sat down with groups that were basically involved with this, to see if we could come up with a solution.

This was not an easy thing at the beginning. We had to sit

there and debate over many things. The organizations that were involved in working out the legislation were the State Health Department, the New Jersey Veterinary Medical Association, and the humane interests, represented basically by the Mid-Atlantic Region of the Humane Society of the United States, Nina Austenberg, the director. We also had involved Lois Stevenson of the *Star-Ledger*, which we felt was very important.

We started back in 1983. By the time the bill came out, it was roughly around 1984. The first bill included indigent people, people who were on public welfare, and would allow them, for a $10 co-payment, to have their animals spayed and neutered. Now, this represented about 15 percent of our population. In 1985 and '86, after looking at this program, we decided to make a couple of amendments to the bill to improve it and include more animals in the program. In the first bill, we had set down certain goals and felt that this was going to be the solution. After the program got started, we felt that there were a lot more ways to go about figuring this out.

I think what I'm going to do is tell you a little bit about how the bill works at present. We don't have a lot of working time with the last bill, since it was 1986 when the amendment was passed to include animals that were adopted from recognized humane shelters and impounding state facilities. The funds themselves come from a $3 payment that's above the licensing fee for people who have unneutered pets.[21] This is the basic fund that we work with. People who are qualified for the program, bring their pets to a veterinarian who is recognized as participating in the program.

Now, as far as veterinarians are concerned—basically, it's a voluntary program, and the State Association will go through and inspect the hospitals. If they pass our hospital inspection, the Health Department then certifies the veterinarians as being qualified to participate in the program.

People can take advantage of the program if they're on some form of public assistance, and there are nine of these that are listed. They have to have some type of state ID, and for a $10 co-payment, the pet is spayed or neutered, and can have all of its vaccinations.

For a $20 co-payment, any animal that is adopted out of a recognized shelter or pound can also have the vaccinations and

## What Part Can Veterinarians Play?

be spayed. Now, the problem with the second part of this, is in terms of the licensed shelters, proving that the animal was adopted from one of these shelters. It was basically up to the Health Department to do this, and there were many, many problems that were involved with working this out.

We also included into the second law that these pets had to be licensed. We're extremely concerned in the State of New Jersey about the control of rabies. We feel that we have an excellent rabies control program. Again, this is cooperation of the Health Department and the veterinary association, and we feel that it's very important that these animals be licensed and also be vaccinated against rabies.

The program itself is statewide. It involves somewhere between 130 and 150 veterinary facilities, which represents somewhere in the neighborhood of around 300 veterinarians that are participating in the program. The number of impounding facilities, I'm not sure, we're still in the process of hearing back from all of these, as you can appreciate. There are an awful lot of these particular animal welfare shelters from one end of the state to the other. Since the passage December 17th, 1986, of the amendment, the program itself has increased, I believe, by 88 percent.

We feel that we are definitely on the way to making a small inroad into this pet overpopulation problem, but again, the big thing that we've found is that the key is co-operation. We all have to set aside our problems that we have with each other, sit down at a table and talk to each other. That's the only way that we're really going to come up with a true solution to this problem. We all have to sit there and talk. In New Jersey, we're very proud of our program, and we feel that it's going to be an extreme success, and this is only one of the solutions.

Since we sat down and began to talk with the organizations, we've come up with solutions to even other problems. For example, pet shop regulations was another problem that we had to look at. We got together and we have some good things here. So again, we feel the key is sitting down and really discussing this problem, because we're all in this business together for the welfare and the well-being of the animals.

**VALUSEK:** Thank you, Dr. Baeder.

Dr. Robert Case

**VALUSEK:** Next, we'll hear from Dr. Case.

**DR. ROBERT CASE:** Thank you very much. It's my pleasure to be here. I welcome this opportunity, and I hope that I can contribute in some way to the success of the program: pet population explosion, pet overpopulation and stray cats and dogs, what part can veterinarians play? They *can* play a part, and they have a role from the standpoint of education. The education of animal owners is the most important part of an animal control program, a program that includes informing animal owners of the need to have their pets "sterilized," "spayed," "neutered," whatever terminology is most appropriate. Now, veterinarians can help educate animal owners through normal veterinarian-client relationships in your private practices. My experience is, though, that most of the people that bring in their animals for veterinary care pretty much have their minds made up that they intend to have their animals spayed or neutered. Now, if we cannot reach the people that need the education because they have decided, or have not decided to bring their animals to a veterinarian, then we're not going to be able to do our job from the standpoint of education. Therefore, we have to reach out as individuals interested and concerned for animal welfare, we have to reach out, and whatever abilities we have from the standpoint of being educators, we have to make that effort.

How can we do this? We can do it by speaking at elementary, secondary school assemblies or programs as guest speakers or for individual school classes. The need to learn about proper veterinary care is present at an early age, and, as we all know, these youngsters have an interest, and basically everybody likes animals. And they're certainly going to be better prepared to be responsible pet owners if they can be educated about what constitutes pet responsibility at as young an age as possible.

In New York State, it is my understanding, and I believe I'm correct, that it is mandated by the education department that schools teach proper care and humane treatment of animals.[22] I don't know how many schools actually involve this in their curriculum, but I would venture to guess that the majority of them do not—not because they're not interested or don't want to, but I presume some of their other mandated items don't get into the curriculum either. But we certainly can, as veterinarians, encourage implementation of this requirement, and we might even have to talk to one or more members of Boards of Education in order to accomplish this. And, of course, we can't just talk about it, we have to be willing to be participants. I'm aware of adult education courses, at least a few instances, where obedience-type work has been taught in those types of classes. Now, obedience and well behaved animals, of course, are part of responsible pet care, but there's a lot more that needs to be explained, as you know, about pet care. And again, my opinion here is that those who I think probably need the education the most are the least likely to sign up for the courses.

Sterilization, spay/neutering of companion animals, of course, is a part of an overall pet population program. Veterinarians expect the practice of veterinary medicine to be performed by veterinarians. I feel most veterinarians are more than willing to participate in a spay and neuter program if they feel their views and opinions have been taken into consideration. No type of animal control program, certainly not a total animal control program, can be successful without the inclusion of a veterinarian. We must try and look at the total picture.

Even though you may think we are doing our part, either by having our own pet spayed or neutered or by being actively involved in some local program, our efforts have not been sufficient. We need not argue as to whether there is a pet population explosion or not. For purposes of frank and meaningful discussion, we only have to conclude that there are too many stray animals, dogs and cats, too many abandoned companion animals, and just too many of these animals are euthanatized. We can't just be satisfied that the euthanasia is humane. We must reduce the numbers.

The numbers euthanatized can be reduced if we, of course, simply increase adoptions.[4] October has been designated as Adopt-A-Dog month. There is a sponsor. The sponsor has made a commitment to over 800 animal shelters. The human-companion animal bond, real or imagined? It's real. Who are we to question a person's qualifications to own an animal? Would they have to answer written questions? That couldn't be done, probably;, we may think it couldn't, because it would decrease the interest in adopting animals, it would discourage people from adopting animals. We would then be guilty of helping to prevent people from experiencing a human-companion animal bond and interfering with their physical as

## Dr. Robert Case

well as their psychological well-being.

The best programs in companion animal population control are those worked out between veterinarians and other interested persons within a community or area. These programs, no matter how good or what the number of animals reported as being spayed or neutered is, have still not resulted in significant reduction in too many communities. Granted, there are some examples where they have been successful.

Responsibility for the cause of the problem? We should be willing to share part of the blame. We should be willing to share in the time and effort required to find a solution. Animal owners that have the resources to pay for procedures should not have those procedures subsidized. However, I support the one-third/one-third/one-third concept of shared fees for adopted animals. And I support starting the process towards the spaying or neutering of younger, immature animals that are adopted. In New York State, in my opinion, municipalities have not done enough to help with educating people regarding responsible pet ownership. That's what part of the licensing fee is supposed to go for. At least the law provides for it.

The problem, however, will not be solved by national or regional or state conferences on the subject if we don't go home and work at the local level. I'd just like to say, as a veterinarian, that I think there is an increased awareness throughout the country from the standpoint of the need for proper animal care. I can best express it from the standpoint of veterinary students themselves. I can recall going to veterinary school when, yes, of course, we were concerned about overt examples of cruelty to animals. But I didn't at that time think about the dogs that we worked on from the standpoint of student surgery, and I certainly know today that veterinary students *do think and are thinking* about this type of situation. And you could say that perhaps the admissions committee didn't do a thorough enough job. However, we all learn by living and being involved, and with this increased awareness, I know that there will be better care provided for animals. This we all can hope for. If we work hard and work together, it will be accomplished.

I'd like to again thank you for the opportunity of speaking here, and I think it's a fine example of your efforts to hear from veterinarians and your willingness to have more of us share in your program. Thank you very much.

[See Update, p. 236]

## What Part Can Veterinarians Play?

**VALUSEK:** Thank you, Dr. Case. Are there any questions now for either of the veterinarians we've just heard from?

**BARBARA CASSIDY:** I'm from the Humane Society of the U.S. and I want to ask Dr. Case to clarify. I want to be sure I understood. Did you say that you did not think that shelters should pre-screen adoptions in any way?

**CASE:** No, I did not mean that, but that's what I said, and I'm glad you asked for clarification. I merely raised it as a question from the standpoint of the interest in adopting out animals: Do we need to think more about what qualifies an individual, if I might use that phrase, to actually own an animal? That's what I'm trying to get at.

**CASSIDY:** That is a strong emphasis that we make, that shelters absolutely must pre-screen so that unsuitable adopters—those that are basically irresponsible owners—don't get animals when we are already euthanizing so many millions.

**CASE:** But from the standpoint of even going further, are we reaching the point where we should be, or is it too much to ask from them that they perhaps take a simple course in ownership responsibility in order to qualify?

**CASSIDY:** It's not too much to ask. Good point. Thank you.

**PAMELA MARSEN:** I wanted to ask Dr. Baeder, in the New Jersey program, what is the ratio between the monies that you're collecting with the licensing differential and what you're expending? Are you taking in a huge surplus, breaking even?

**BAEDER:** What I would like to do is refer the answer to that to Dr. Sorhage from the Health Department. Since they're really responsible for the money, they do keep track of the statistics there. Possibly, she could give you a better answer on that.

**DR. FAY SORHAGE:** I'm the State Public Health Veterinarian of New Jersey. I don't directly run the spay/neuter program, but I share the office with the person who does, Bob Monyer. So I know a little about it, and I serve as the veterinary consultant for the program. As far as the financial aspects of the program are concerned, the first couple of years that it was in operation, it took some time to get the publicity out about it, to get people into the program, so there was a large amount of money accruing, quite rapidly, that wasn't being spent. That's one of the reasons that the amendment was passed, to also include adopted animals from shelters into the program to get spayed or neutered, along with animals belonging to persons who were on public assistance programs, so with this added to the program, we expect the funds to be used up a little more rapidly. Hopefully, we'll make a break-even point at some point. As far as the future of it goes, if theoretically it works right and all the animals get spayed, there will be no more funds going into the program, because everybody buying a dog license, their dogs will be spayed or neutered.[21] So, that could be the end result, that the program could end, but I kind of doubt that there will never be any unaltered animals.

**MARSEN:** And I also have a question for Dr. Case. As I mentioned at a presentation yesterday, I was very concerned two years ago at the HSUS convention in Chicago to hear the incoming president of the American Veterinary Medical Association and his *confrères* say that they weren't concerned about the overpopulation. They didn't think there necessarily was much of an overpopulation and furthermore, they could even stand a slight increase in some areas. I hear people many times say: "Of course, veterinarians aren't going to be for extensive spay/neuter, because then there will be fewer animals for them to treat and less money for them to make." Could you please tell me your perspective on that?

**CASE:** I think you understand that I represent myself here, today. But anyhow, I'm glad to try to answer your question. I think analysis has to be done in each and every community. If I use my own, for example, I've had to conclude that a very good animal control program exists. I'm the resident of a town between two sizable cities, and they have, I think, a fairly ample staff and good vehicles. Animals that appear to be injured or hurt in some manner are taken to veterinarians

## What Part Can Veterinarians Play?

including the fairly recently opened emergency clinic. I can't speak for my profession as a whole, but if there are problems in a community that result from animals that are strays and if we can definitely point out that there's an excess population, I really don't think that veterinarians are concerned that too many are going to have the operation performed and that is going to reduce the numbers out there from the standpoint of work for veterinarians.

Statistics now are showing that more and more people have cats, and there's more than one reason for this, of course. One might be in situations in multi-family dwellings in maybe smaller communities where it might or might not be easier to speak to the landlord and it can be explained to them that this nice feline does not represent the destructive potential that there would be with, for example, a Great Dane here, and so on and so forth, And so I think we have to again look at the total picture. There's an obvious interest in people having pet animals, and I don't think we're going to reduce the supply from that standpoint. If the interest is there, then we'll work out something. We're just not going to spay too many animals.

I hope I've answered your question.

**GRETCHEN WYLER:** I am enormously impressed with what you were talking about in New Jersey. I would like to ask a couple of questions. First of all, was this legislation motivated, lobbied for or fought for by humane groups? And would you identify which ones started it?

**DR. BAEDER:** The lobbying really came from a combination of the veterinary group and—

**WYLER:** Who went to whom?

**BAEDER:** Well, it's kind of hard to say who was the one who really initiated this.

**WYLER:** It doesn't sound to me like something the veterinarians would have started.

**BAEDER:** Well, we did. We actually had a liaison, back several years ago. He would have to be really considered the driving force of this; this is Dr. Sidney Rosenberg. He and Nina Austenberg, from the Mid-Atlantic region of the Humane Society of the United States, were the two who really got together and started sitting down—he as representing our veterinary association, and she as representing the Humane Society. This is how the initial discussion took place. The strongest lobbyist for this was one of our own lobbyists.

**WYLER:** You see what can happen when we do sit down. Traditionally, you know, we have great trouble.

**BAEDER:** Exactly. Well, that last point was the one that I was trying to make.

**WYLER:** And we're going to continue to because of your position on pound seizure, because of your position on trapping. As an association—I don't mean as a veterinarian, but as an association. So I'm very impressed with this, and I hope it bodes well for the future. What do the veterinarians around the country think of your program, the AVMA, say?

**BAEDER:** Well, right now, everybody's looking at it, because it's the first of its kind. It looks to be successful and we certainly hope it's successful. And once it comes about, I think that you may begin to see more of these people sitting down at the table. It's amazing in terms of what we can do when we actually sit down and talk to each other about this, because we have to remember that there's one common thing that we're all here for, and that's the welfare of the animals, the health and the well-being.

**WYLER:** Not always. That has not been our history, and that's why I think this is wonderful, I think it bodes well. Obviously, I'm a little worried about it because of my history with the Association in the past, on other issues: food animals, fur animals. And when it comes to the pet animals, I certainly think that should be the first place that we try to share this common ground. And I know I represent a lot of people here who would commend this affiliation. We've been resistant to it because of the history that we've had with the veterinary associations on things like food animals and fur animals and other kinds of animal welfare, animal rights, issues. I think this is remarkable in its cooperative effort between the two, and I would applaud you for that. I am surprised that it was so successful with your association. I thought perhaps you were someone who was just acting independently. But you're saying the New Jersey Veterinary Association absolutely lobbied for this, your lobbyists?

**DR. BAEDER:** Yes.

**WYLER:** That's wonderful. How do you think it portends for the rest of the country, because it could be a major thing?

**BAEDER:** Well, we have high hopes for it. We're extremely proud of this program, and we just hope that it sends up a light to the rest of the profession.

**WYLER:** And I hope that you would aggressively be activistic and get around the country, because I think it's major, so I applaud you.

**DOROTHY DENGEL:** I am a lay person from Kings Park, Long Island, and when Dr. Case was speaking of the ways and means of educating people, he mentioned the assemblies for the children, adult programs, spay/neuter program, and then he made a comment: Very often, people that need it the most, either the poor or the less educated people, are the ones that don't come to adult education courses. I'm just wondering: Do we take enough advantage of volunteers, either within shelter programs or education programs? Not that they have to be teachers. Can there be given out, not necessarily wordy things about the rearing of animals, but even pictures if we have to, when people are non-readers? Have we gone to where the people are, rather than waiting for them to come to us or taking advantage of, say, veterinary situations when they come to us? Do we go to flea markets? I'm struck by the number of poor that go to Aqueduct flea markets. Do we set up a table? I know this involves money. Do what we do at some big conferences, professional conferences, where they have a little movie going on all at the same time while it's a loud voice possibly saying: "This is what you do with your puppy. This is Johnny. This is Johnny taking care of his puppy." Do we go to thrift stores, do we stand outside of thrift stores where the poor go? Do we take advantage of senior citizen groups who are looking for things to do for the community? Maybe organizations and associations could say: "Hey, Port Jefferson, Long Island, singing group, would you give out these pamphlets door to door in such and such a neighborhood?" Where people are not able to take veterinary trips. They are letting their animals be, because they can't pay for it. Or they don't know how to go about getting their animal taken care of. I'm just saying, people like myself, lay people, who may not feel inclined to teach or don't think they have enough information up here about teaching would surely be able to give out papers or stand at a flea market table. To do things like that. That was just my statement, it's not really a question.

**VALUSEK:** Thank you. Is there perhaps anybody in the audience who is doing something of that nature, a grassroots sort of thing? Has anybody taken advantage of the local senior citizens or groups? Is there anyone who would like to address the comments?

**SHERI TRAINER (Bide-A-Wee Home):** We have expanded a new educational services department, and rather than answer all the points that you brought up, I'd like to defer that to the afternoon session, because I will be addressing a lot of these issues.

**NINA AUSTENBERG (HUMANE SOCIETY OF THE UNITED STATES):** This isn't a question, it's just a statement: The New Jersey Veterinary Medical Association has helped us on a number of issues, and I thought maybe I should say that. I'm not a spokesperson for the New Jersey Veterinary Medical Association, but we worked on the banning of the sale of wild birds, the pets-in-housing legislation, the certification of animal control officers, and the total ban of the steel-jaw leghold trap. And in all of those, the New Jersey Veterinary Medical Association has worked with us and sent their lobbies to testify, so they deserve a pat on the back.

**ANN FREE:** I don't believe that I got it down quite right. Could you give us a rundown on the financing of the New Jersey program; what the veterinarians' take-home is?

**BAEDER:** Okay, the basic fee is the $3 that's above the licensing fee. The veterinarians themselves take 80 percent of what is considered their customary fee. Now, what basically has to happen is that we send in our fee schedules to the Health Department, and on a county-wide basis, the Health Department looks at these fees and then comes back and okays them, to say that: "Okay, within your county, within your region, your fees are acceptable," and they reimburse us 80 percent.

**FREE:** What's your reasonable fee for a medium-sized female dog?

**BAEDER:** Well, again, it varies, but to be honest with you, I can't even remember what our fee here is, but it probably runs around $65, around $65, $75.

**FREE:** Well, in short, you're getting, let's say, $50.

**BAEDER:** Yes, somewhere between $50 and $60.

**FREE:** And $10 comes from the individual.

**BAEDER:** No, the $10 co-payment actually goes back to the State, that helped the party. I believe that's for the paperwork and everything else. There's a $10 co-payment on the one program and $20 co-payment on the other.

**FREE:** Right. Where is that money coming from? Your pay?

**BAEDER:** Our pay comes from the $3 "excise tax," or however you want to look at it.

**FREE:** I'm quite surprised that there are that many three dollars, I mean, that that many animals—

**BAEDER:** Well, there are a lot.[21] I believe that somewhat after the first bill was signed, they had in that fund, I believe it was close to $2 million.

**FREE:** Mainly males, I guess.

**BAEDER:** Well, there are a lot of females too, in fact, in the program. It's surprising how many unaltered pets there really are out there.

**FREE:** Well, it's good you smoked that one out. Thank you very much for reviewing this with us.

**ELAINE BIRKHOLZ (Special Projects Manager at the MSPCA in Boston):** Does your program, Dr. Baeder, require pre-sterilization inoculations and, if so, is there any kind of a discount program for those?

**BAEDER:** The inoculations themselves are basically included in the program.

**BIRKHOLZ:** Those are given when the sterilization—

**BAEDER:** Yes, at the time of the neutering we include the standard DHLPP and the rabies inoculation. I could add that, when the vaccinations are included, the veterinarians are reimbursed an additional $10 fee for the vaccinations.

# Neutering Sexually Immature Animals

**MARJORIE ANCHEL:** I would like to introduce our next speaker, Dr. Leo Lieberman.

Dr. Lieberman is a veterinarian who is now practicing in Florida, and did practice in Connecticut. He is interested in the neutering of immature puppies and kittens, a very important subject for us. I first got an inkling of what Dr. Lieberman was doing, several years ago, through a letter in the Journal of the AVMA, which spoke about early spay/neuter. Recently, I got in touch with Dr. Lieberman, and found that he had just published a full paper on this subject. We now have the pleasure of hearing him talk about early spay/neuter, a possibility of immense importance for the whole question of overpopulation.

**DR. LEO LIEBERMAN:** The lecture will be prefaced by a videotape. This videotape is in need of editing, but it's very salient to the point.

**ANNOUNCER [videotape]:**

> *Everybody at the Humane Society recognizes that pet population control is a real problem. There is a huge void in the programs that exist today. Society recognizes the value of animals to youngsters and to the elderly, as well as their importance to family life. However, animal shelters and humane societies are confronted with the severe problem of being inundated with unwanted puppies and kittens. The National Animal Control Association tells us that it costs between $60 million and $100 million a year*

to control the problems associated with the lovable but unwanted kittens and unwanted puppies. Only one in ten finds a home.

What is the crux of the real problem? The desire to reproduce is basic to all animals, but when one animal gets loose to reproduce at will over a period of three or four years, there will be approximately 4,000 unwanted puppies and kittens born. Four thousand animals that wind up the responsibility of animal shelters and humane societies. They must do something with these animals. If only one out of ten is adopted, euthanasia seems to be the prevalent answer. But another answer is to make the animal more adoptable and at the same time, control the population. Today, modern veterinary medicine has the capability and the drugs to anesthetize and perform surgery on very young animals. The concept of performing surgery on animals at eight weeks of age prior to adoption in the home is not a new idea, and it certainly deserves serious consideration. Young puppies and kittens can be safely neutered at eight weeks of age. Of course, this can be a real plus in the adoption program.

If puppies and kittens are healthy enough to be adopted, they are sturdy enough for surgery. The bottom line here is that these animals can now go into homes, be real rewarding for the owner, and not create more puppies and kittens.

What are the advantages of such a program? Improved quality of life? Absolutely. The animals become more lovable, more playful, stay at home more readily, adapt themselves very well to family life, and have an extended life expectancy. For the shelters, the animal is more adoptable. There is no need for the adopter to come back in six months, and fewer animals are returned unwanted.

With early neutering, the reduction of unwanted animals is significant. If you recognize the fact that today the second largest cause of death to pet animals is the automobile, you must also recognize that the first largest cause of death in pet animals is shelter euthanasia. There is no way that an animal can argue with an automobile and win. Sex drives them to wander, and that is a most serious consideration for this program.

Another added benefit to this early neutering program would be an opportunity to implement an identification system. When the animal is under anesthesia, a tattoo can be applied or a microchip can be inserted. Thus, strayed or lost animals could be identified and returned to their owners.

One of the concerns of people regarding this program is the prospect of overweight pets. Realizing that owners love to feed their pets, there seems to be no sign of strict correlation between overweight neutered animals and non-neutered animals. It is the people who feed the animals who cause the problem.

Another serious consideration one must think of is feline urological syndrome or FUS. Today, science tells us that there is no correlation between the age at which a cat is castrated and the occurrence of feline urological syndrome. The urinary problem of incontinence is similarly unrelated to the age at which the animal is neutered.

The bottom line is that those animals who are neutered early in life make better pets and are more likely to be adopted. They are more playful, stay at home, and are more responsive. The owners become more responsible as well. Surveys tell us that owners who have pets who have been neutered early are more satisfied, find the pets more adaptable, better-behaved and more healthy.

One factor of real consideration is that there is real reduction in the number of unwanted animals euthanized. One shelter reports a 63 percent reduction. Pet population control and very early age neutering are compatible, and complement each other. It certainly should be given serious consideration. You know, a pet is very special friend.

**LIEBERMAN:** Good afternoon, ladies and gentlemen. Thank you, President Anchel, members of the New York State

### Dr. Leo Lieberman

Humane Association, fellow veterinarians and guests. It's a pleasure to be here to propound on the merits of neutering puppies and kittens at eight weeks of age, the time of adoption. The use of word "neuter" here means spaying and castrating.

Everywhere you hear that it's raining cats and dogs. More sterilizations are needed to stop the storm. No one can disagree that a pet population explosion has occurred during the past several years throughout the United States. The purpose of this lecture is to have veterinary surgeons persuaded to do this surgery at eight weeks of age, before adoption. Also, shelter administrators must be persuaded to implement this concept.

To assist in validating this concept, records and follow-up studies must be made available toward publication. Let's all work toward the truth and have notions and opinions put aside. Breeders of registered dogs and cats have expressed interest in this concept as an alternative to selling puppies without papers. Cute little kittens and adorable puppies are most adoptable at this age. They are resilient and tough, and safe surgery can be done, and should be done before adoption. Neuter them.

The usual and legal age for sale and adoption of puppies is eight weeks. This is a reasonable requirement. If these animals are suitable for adoption, they certainly are sturdy enough for surgery. Let one escape sterilization to have a litter of six, three pairs, and if each of these pairs should have a litter of six, and if their offspring would reproduce in the same manner, in four years, there would be approximately 4,000 pairs or more from this one mating. Ninety percent are unwanted, unneeded, and society has no place for them. No wonder it's called a pet population explosion.

The National Animal Control Association tells us that up to 70 percent of animals adopted through shelters and humane societies fail to fulfill the contract of sterilization. One shelter that is required by law to enforce this contract, and takes payment at the time of adoption, still has 30 percent failure rate, despite excellent efforts. Another shelter, in their efforts to solve this dilemma, set a policy in place in which only dogs and cats ready for surgery, that is, at six months of age, were available for adoption. In other words, all puppies and kittens were euthanized. *That's not right.*

There are many precedents and many, many advantages to the concept of neuter surgery at or before adoption. Let's look into them: The American Veterinary Medical Association Animal Welfare Committee in 1985 advocated surgery on farm animals as early as possible. This includes castrating, spaying, dehorning the cattle, swine, sheep and goats. In pigs, it's usually three to five days after birth. In calves, it's usually done within a week after birth. In Australia, colts are castrated at age ten days. In China, young female pigs are spayed at age two months. I have recently found that the ferret is spayed and descented at age six weeks, routinely. There is enough evidence available now that dogs and cats should be added to this early surgery list.

The tradition of doing neuter surgery at six months on dogs and cats started before the 1920s. At that time, there was no suitable anesthetic for cats, and the anesthesia for dogs was less than adequate. The hazards of surgery put the life of the patient at risk. The risk was reduced in a mature animal, and so the conventional age was selected. The tradition grew and grew and grew. It is very difficult to change tradition on *any* subject.

In the intervening years, veterinary surgical skills have kept pace with the improvements of surgery on people, with much of the same equipment and drugs available to the veterinary practitioner. This has increased the sophistication and skill, improved the services to the public and the safety to the patient. These are some of the common drugs available to veterinarians for anesthesia: Ketaset and acepromazine are the popular anesthetics for these very young animals. There is also metaphane used in the anesthetic machine.

The skill of veterinary surgery has reached a place where surgery on parakeets is not unusual. "Pocket pets" such as rats, hamsters and gerbils are often subjected to successful surgery. Purveyors of laboratory animals regularly offer surgical services on mice, rats, hamsters and guinea pigs, including surgery on adrenal and pituitary glands. Think just a little bit, if surgery is being done regularly on these tiny creatures, it certainly is feasible and practical to neuter puppies and kittens before adoption at eight weeks of age. Such a program is not only beneficial to the animal shelter, to society, but to the animals themselves.

The different sizes of animals makes a difference in the difficulty of surgery, and the size of the fee. This is reflected in the sizes of the uterus. You can see from the slide: The one on the left is a juvenile; the one on the right is from an adult. Certainly, one can appreciate the need to elect surgery as early as possible.

One could say that neutering at eight weeks of age is empiric, that is, lacking scientific basis. That's true. Actually, I have reports on more than 10,000 early neuter surgeries. You could say the same about conventional age for neuter surgery. Six months of age is also empiric. There is no scientific basis. A search through the scientific literature by many people has developed *no* information on this subject. In a recent conversation with the librarian of the National Agricultural Library, where there are 2.6 million volumes plus 26,000 periodicals regularly received, I learned that the librarian has spent two years trying to develop a literature search on this subject. He recently told me he's going to send me a letter that there is no information on this subject.

The origin of neutering at two months of age is not mine. In 1925, Dr. Flynn of Chicago described and published what was then called the sutureless spay. It was recommended to be done on two- to three-month-old puppies and kittens. It did not become popular. I suspect that anesthetic skills had not progressed adequately. In 1982, I reported in the Journal of the AVMA that I had done 100 puppies and kittens and suggested that it be considered by other practitioners. It was hardly noticed until your president found it and then found *me* in Florida.

Chemicals and other methods have been offered to facilitate and reduce the cost of neutering dogs and cats. They all have been found wanting or have been withdrawn by government order. There is a recent offering of chemical castration by injection of the testicle. I have had difficulty with this procedure unless anesthesia is used. If anesthesia is used, it is easier to do the surgery. Believe me, I have done both.

There is a large number of advantages to the shelters for eight-week-old neutering. If the animal is healthy enough to be adopted, it is sturdy enough for surgery. The adopting owner has no choice. And the price is right. No recall is needed. You all know that people move, refuse, object to surgery, or the animal is lost. Early neutering reduces the paperwork. The pet adapts to the family more quickly; thus, fewer animals are returned to the shelter. One of your members published a paper: "A Study of the Human-Animal Bonds That Fail"—in other words, the reasons for the returned adoptee. There is a significant number of animals that fall into this category, and I believe the numbers can be reduced with very early neuter programs. The puppies and kittens will become more adaptable, more adoptable. Definitely, "a better product" will be created, and this will create more demand and increase adoptions.

This also gives an opportunity to clinically effect an identification program. Under anesthesia, it is a very simple procedure to apply a tattoo or to insert a microchip.[5] I personally know of two facilities where this is practiced. One has a computer for the quick retrieval of tattoo numbers to identify the owner for the return of the stray. In another facility, the owner of the stray is identified and summonsed to court. It also facilitates license renewal and perhaps licensing at the point of sale. I have in hand a tattoo pliers made by the Ketchum Company of Canada and supplied to me by the U.S. pet registration system that offers nationwide service. Here five digits are small enough to fit in a cat's ear. Other tattoo machines and other tattoo identification organizations are also available, as is the choice of colors of ink. With five digits, 100,000 animals can be identified, and if you add the letters of the alphabet in each of the five places, my mathematician friend tells me this amounts to 100,000 plus 26 to the fifth power. Millions can be identified with a simple little machine like this. This identification program reduces the number of stray animals.

There would be fewer unwanted litters. There would be fewer euthanasias. One shelter reports 63 percent reduction of euthanasias in seven years. Expenses to the shelter would be markedly reduced. The advantages to the pet are even more striking. The surgery is safe, there is less shock, there is quick recovery, and very rapid healing. There is an extended life expectancy, and the quality of life is improved, not only for the pet but for the owner. There is less mammary cancer; the glands are not stimulated to function. There is no pyometra; the uterus has been removed. There is no prostate disease; there would be no androgen secreted. The incidence of perineal hernias would be reduced. There would be an increased owner responsibility toward a less aggressive, more lovable pet who was less likely to bite people or other animals. This is very important to the animal control personnel, who have to investigate the incidence of dog bites. The Vancouver report shows that 81 percent of dog bite incidents were caused by uncastrated animals.

Dr. Leo Lieberman

The second most common cause of death to pet animals, since infectious diseases have come under control, is the automobile. The first cause of death to pets in this country is euthanasia. Neuter-at-adoption, my proposal, reduces both.

A recent study in Baltimore of 5,000 cats that were found killed on the city streets indicates that 90 percent of these casualties were sexually intact individuals. Sixty-three percent of these were intact males. This tells us that if these animals had been neutered, they probably would not have been crossing the street in the first place. This gives some food for thought. These early neutered animals stay at home, are playful, lovable, mild-mannered, what I like to describe as the perennial sophomore attitude.

The advantages for the veterinary surgeon are very noticeable. The anesthetic is simple: Ketaset, acepromazine or other. The surgeon at the University of Florida, who's working on a project involving six-week-old bitch puppies, used nitrous oxide for the anesthesia and described the procedure as, "a piece of cake." He had never done it before. He didn't realize it was so easy. The puppy was eating in less than an hour. Recovery is rapid. The patient is ambulatory in a short time. A small incision makes for rapid wound healing. There is very little visceral fat, which simplifies surgery. The ovarian ligament is very elastic and easily manipulated. There is reduced time for the procedure and reduced cost. Anxiety is reduced for the surgeon and the patient. These very young animals are resilient and tolerate the procedure very well.

The potential results of this early neuter are: reduced number of unwanted litters, reduced stray animals, reduced euthanasia, increased owner responsibility, improved humane treatment of animals, and improved quality of life for the animals and the owners.

Some questions have been asked about disadvantages or undesirable aspects of early age neuter: What are the long-term effects?

One long-term effect that's been reported is that the bones of the legs, mainly the radius and the ulna, grow slightly longer. This is measurable when the subject is compared with control animals. However, it is hardly noticeable, and there have not been any objections from owners who have these animals.

Behavioral changes that have been noticed have all been desirable changes, as referred to previously.

The surgery has been found to be less hazardous than usual.

Humane organizations are very wary about using very early neuter surgery in their own facilities or recommending it, because of their concern about possible criticism from other humane organizations. I know that this sounds peculiar, but it is true. It is time now that this concept be put on the table for open discussion among yourselves and your veterinarians. Please do not dismiss it out of hand.

There are many veterinarians and many cat-owning people who honestly are concerned that two-month-old neutering predisposes the male feline to FUS (feline urethral syndrome). This slide shows a schematic of the urethra plugs in the distal end of the urethra and gives you an idea what the problem is. Lon Lewis, in his 1987 book, *Small Animal Clinical Nutrition #3* categorically states that all sexes are equally susceptible to FUS, that is: intact males, intact females, neutered males and spayed females. The time or age of surgery is not related to the incidence of disease. He has references, quotes his own and other studies to substantiate the statement. In my own experience, I agree. Low magnesium diets for prevention are readily available. Carl Osborne, the urologist from the University of Minnesota, is of the same opinion.

This slide shows a schematic of the formation of struvite crystals. The ammonium, the phosphate and the magnesium combine to form the struvite. The struvite crystals aggregate and have a gummy substance that forms into stones and uroliths. Remove the magnesium, and the struvite crystals do not form. Where they do form, they form throughout the urinary tract. And this bladder stone indicates the formation of stones in the bladder; and notice the cystitis, the inflammation and background there. The alkalinity causes the inflammation that you see. Urinary tract inflammation is a common precursor of FUS. The alkalinity of urine enhances the formation of struvite. One manufacturer of kitty litter is planning a product that will change color when the pH is greater than 6.7. This color warning will give ample opportunity to start treatment promptly.

There is a misconception that draws a direct relationship between neutering and obesity. The practicing veterinarian is having difficulty solving the problem of obesity. He finds it in intact as well as neutered animals. He also finds that owners receive great satisfaction seeing their pet eat well, and the owners pander to it. Lack of exercise is also a very big factor. Looking at the TV ads and the store shelves will quickly explain some of this pleasure, plus these soft, moist burgers use sugar as a preservative. This increases the caloric value, and the sugar is reported to be addictive. There is only one controlled study of which I am aware in which weight records were accurately maintained. Sixty shepherd bitches were in the study. Twenty were spayed; twenty were intact; and twenty had ovarian implants. All were fed, worked and housed in an identical manner. At the end of a year, there was no significant difference in weight—one-half pound between the heaviest and the lightest. Do not blame the surgery for obesity.

Neutering, particularly early age neutering, is reputed to influence an increased incidence of incontinence, a lack of controlled urination. That opinion has not been validated. Each case of incontinence requires its own diagnostic work-up, which can be very complicated. Incontinence is classified as neurogenic or non–neurogenic. This slide shows the complicated nerve system that influences the voluntary control of urination. Each nerve, each area, can be affected by trauma or disease and be very difficult to diagnose. Genetic abnormalities are also factored in. After an accurate diagnosis, treatments give reasonable results. In this list of treatments is estrogen, and it often works well. Just how and why it works is not understood. However, some have speculated that because it works indicates a deficiency in the hormone. That theory has *not* been verified.

Non-neurogenic incontinence has a variety of causes. More than 50 percent of this category is caused by urinary tract infections. The bacteria cause an alkaline urine, which causes a burning sensation and frequent urination. This burning sensation is very similar to the burning sensation that you get when you get soap in your eyes. Tumors and stones are also a factor in urinary incontinence, as is the incoordination of the muscles of the bladder and the urethra. Treatment of cystitis works well usually but there is a tendency for the cystitis to recur. Careful, continuous monitoring is necessary. Last, but not least, the behavior problems. They must be considered and eliminated as the cause of incontinence. These pets get mad at their owners. The differential diagnosis *always* includes behavior, and treatment depends upon the diagnosis.

Prevention of incontinence should receive strong consideration. And this can be readily accomplished at the annual physical examination by a complete urinalysis, including pH, specific gravity and sediment.

Treatment depends upon the findings. Urinary incontinence occurs often in older animals, male and female, and it is not well understood, even in people. This paper by Thrustfield, of Edinburgh, reports statistics of a relationship between incontinence and spaying in dogs more than six months of age. They are interesting statistics. The total incidence of incontinence is small, 76 incontinent bitches out of 3,200. Of these, 46 were spayed versus 30 intact, a significant statistical difference. However, no diagnosis is reported in any of these cases. Correspondence with Thrustfield indicates that this is a preliminary report and he is continuing his work in cooperation with the University of Liverpool.

In addition to my own experience with very early age neutering, there are a few places that I know where this procedure is being done regularly. There must be others, and I'd like to know about them. A Florida SPCA shelter regularly neuters all pups and kittens and has been doing it for more than five years with no complaints except one: A member of another local humane society felt repulsed by surgery on such tiny creatures. With some effort, she was placated. There have been no other complaints. They report 1,800 animals, eight to twelve weeks old, neutered and adopted. All have had favorable reactions from the owners. No retrospective study has been done. They also report that, for ten years, all males have been neutered before leaving the shelter. There have been no problems.

Medford, Oregon, SPCA shelter reports with enthusiasm more than 10,000 pups and kittens neutered since 1973. There was some concern about rumors of long-term undesirable effects. Repeated inquiry to many sources found no information available. So the shelter funded a retrospective study on their own. The owners of 200 male and female dogs neutered at six to twelve weeks of age, now three to twelve years old, were sent questionnaires. For comparison, 200 owners of dogs neutered after six months of age or not neutered, now three to twelve years old, were sent the same set of questions. The answers were tabulated in these charts. Notice the favorable reaction of the owners. The crosshatch is the early neutered animals. The owners like the early neutering. The dogs appear to be more healthy, less aggressive, more intelligent. Also notice: Weight control favors those animals neutered at an early age. That was the female list. In the chart for males, we find similar reactions, with small differences in percentages. The owners are generally pleased.

A similar questionnaire was sent to the owners of 120 cats neutered at six to twelve weeks of age, now three to twelve years later. Favorable replies were received from all owners. Three individual cats were reported who occasionally sprayed urine. There was no control group.

This chart shows a dramatic drop in euthanasia statistics, which the shelter attributes to the very early neuter program. Notice that in 1973, there were 14,000; 1979, there were 9,000. This is a considerable drop in euthanasia rates.

The purpose of this lecture is to have veterinary surgeons persuaded to do this surgery at eight weeks of age, before adoption. Also, shelter administrators must be persuaded to implement this concept. To assist in validating this concept, records and follow-up studies must be available toward publication. I want to publicly thank Dr. Thomas Lane of the University of Florida for his assistance in preparing this material. I also want to thank Gretchen Wyler for having suggested this idea to me and for the encouragement to continue. This material was published on the first of September in the Journal of the AVMA.

If there are questions, now is the time.


**DR. WOLFGANG JÖCHLE:** I think we should not kid ourselves: Castration generally alters metabolic rates in the body. There is a different metabolism with or without sex coordinates, and there is less protein produced. There is more carbohydrate produced and more fat produced out of the food. It is our responsibility actually to take care of this, and by keeping the dog active, or the cat, and not feeding too much, we can certainly avoid obesity. But there *are* differences in response to neutering and spaying. I think we should not forget this.

**LIEBERMAN:** I understand this, and I believe that most people have given this aspect of the problem more serious consideration than they should. By controlling the diet, there is no doubt you can control the weight.

## Neutering Sexually Immature Animals

**JÖCHLE:** Absolutely.

**LIEBERMAN:** I am glad we agree.

**JÖCHLE:** And there is some information available about the effect of estrogens in incontinence. It has nothing to do with the fact that a hormone is missing. It has something to do with the fact that very often or in a certain percentage, the stump of the cervix adheres to the bladder, and that causes problems. When estrogen is given, it softens the tissue and softens the adhesion, and therefore, it can temporarily make the problem disappear. Take the estrogen effect away, and the whole problem comes back. And this is being shown by research in Europe in which you go in by surgery and remove that adhesion. Then you have the incontinence problem in those cases solved, and you don't need estrogen.

**LIEBERMAN:** That material is available in this country too, and I debated a great deal about including more information on this. But some of this is so sophisticated and difficult to understand that I decided not to include it. And you, being knowledgeable, can appreciate where here the talk is being directed toward basically a lay audience, I just did not want to get into those complications. What you're saying is true, but I want to reiterate that the effect of the estrogens is not understood, and I think you'll agree to that too.

**VOICE:** We have an animal rights coalition in the Bronx. We get a certain problem that I feel everyone here does. Perhaps we get it a little more. This is with human males who do not want their male dogs castrated. It seems to occur more with lower educated men. It is not funny; we know it's not funny. We get tremendous resistance, especially not only from lower educated men but from Hispanic men. We have many Hispanic men in the Bronx, and this a tremendous insult they feel if you persist with this. And I would like you to please tell us how to handle this problem.

**LIEBERMAN:** I'm sorry. I have no direct answer to that. However, if you notice in the list of one of my slides, anthropomorphism was listed, and there I hesitated to get into a discussion of that factor. And, really, I feel that this is not a veterinarian's problem, but I know that it is a problem, and it's much greater than what you have described.

**VOICE:** I just would like to suggest that this person suggest to these people if they are against castration, vasectomy instead. I have dealt with them and understand.

**GRETCHEN WYLER:** I would just like to ask one question. I regret that the only question so far on this subject, was a negative. I applaud your work, and I know there is nothing in this for you except, I would like to think, thanks from our community for trying to reduce the pet overpopulation. So I would like to extend my thanks. I would like to hope that I am joined by many in the room. I would like to know, since you have had one query that was, as I say, rather negative, what do you think the response of the veterinary community would be, or what has it been so far since the *Journal* published your article. What has the response been from your colleagues?

**LIEBERMAN:** The response has been mixed. Just two days ago (mind you, this article came out on the first of September) I received a telephone call from California. This veterinarian was most anxious to get some of the intricate details and was anxious to try this. And he was so enthusiastic that he was proposing that I get invited out there to speak to the veterinary association. By the way, I've been away from Florida for a few weeks, and the address on the paper is Florida, and I was amazed he was able to find me in Connecticut, but he did. But there may have been other inquiries in Florida, so I really don't know.

When this was presented to a local group, there was strong opposition on the basis of economics. And my reply to that was that with 70 percent of these animals now not getting surgery, we can anticipate more than double the market.[23] I wasn't about to get into an argument about this. However, one cooperating colleague wanted to go ahead and undertake this, but because he knew of the opposition, he suggested this be passed, at least for information, in front of the veterinary board, which was done. The correspondence that took place was quite extensive, and ultimately they gave neither yes nor no. They put it in the same category as acupuncture, and they weren't against it, but they were not going to endorse it—which is all that I asked them to do anyway.

**WYLER:** The only reason I wanted to ask the question was I realized that you have put together this program just for the NYSHA conference, and I know that for ten years you've been trying to promote this—for nothing in your pocket, but for something you thought would help the pet overpopulation problem. And I would hope that many people in the room listen to you. . . .

**LIEBERMAN:** Thank you very much.

**[See Updates, pp. 236-240]**

# CHEMOSTERILANTS

**VALUSEK:** I have the pleasure of introducing Dr. Wolfgang Jöchle, who will talk about chemosterilants. Dr. Jöchle is the president of Wolfgang Jöchle Associates, located in Denville, New Jersey. He's an expert in the reproduction of domestic animals and in pet population control; these are subjects on which he has published and lectured internationally. He's written for more than 200 publications, and he's the publisher and author of *Animal Reproduction Report* and *Animal Health Newsletter*.

**DR. WOLFGANG JÖCHLE:** This presentation deals with non-surgical methods available for the control of reproductive functions in dogs and cats. Physical, chemical and pharmacological methods exist for the reversible or irreversible control of reproductive functions in mature and immature male and female dogs and cats. Not all of them are equally well-suited for use in animals kept as pets, or in stray or feral animals. But together they provide a formidable arsenal of options for responsible pet owners not inclined toward surgical procedures, and for the control of stray or feral animals.

In general, some animal owners resist, consciously or unconsciously, any attempt to see reproduction curtailed. Reproducing animals are to them surrogates for their own children, which they do not want, or cannot have. For others, "taking sex away" is incompatible with animal welfare or even animal rights. A special case is the bond between the male pet owner and the entire male dog or cat, which often prevents neutering of these animals: the owner enjoys seeing an animal do what society does not allow him to do. And there is a very simple answer to this: Try to get the lady of the house to come to the veterinarian. She has rarely objections against having these animals neutered.

The topic of this presentation invites a global view and predictions for the future. A global view reveals a striking difference between the situation in North America and all other developed countries. While in North America, surgical solutions to pet population control clearly dominate, they are of only secondary importance worldwide. On the other hand, chemical/pharmaceutical, methods, used almost exclusively abroad, have not found their way into North America's veterinary offices.

There are good reasons for this and they are buried in the past. Since this situation provides for a good lesson for the subject on hand, their resurrection seems warranted:

About twenty-five years ago, the "pill," i.e., orally effective progestins became available for anticonception in the human female. Simultaneously it was shown that the same progestins could be used to control reproduction in dogs and cats and other domestic animals, equally well. Pills for dogs and cats became available in Europe.

In North America, an injectable product was developed, which formed a drug depot in the body and prevented heat for four to six months. Unknowingly, the dose chosen was, in retrospect, much too high; the knowledge of the basic elements of the reproductive cycle in the bitch was still insufficient to safely use such a product. Undesirable side effects abounded as a consequence, resulting in an epidemic of pyometra. With it, many bitches lost their reproductive capabilities, since surgical ovariohysterectomy was the only solution. And when surgical intervention did not come in time, they lost their lives. The veterinary profession in North America recoiled from the use of hormones for pet population control. Since at about this time simple methods for good anesthesia became generally available, a tradition of surgical neutering methods became established.

Abroad, the same hormonal preparation, which was used initially with such disastrous results, is since more than two decades the most widely used product for estrus prevention in millions of bitches. The reasons for this are: only a quarter or a fifth of the dose recommended initially in North America is used. And increasing insights into the very peculiar, almost unusual conditions of the bitch's cycle have been needed in order to prevent estrus safely, but not to suppress clinically evident proestrus or estrus. Using such products in animals already in heat invites serious complications. In addition, anti-

## Chemosterilants

androgenic products are used abroad, which reversibly and temporarily control sex drive and mating behavior in male dogs and cats, and with it the widely seen, obnoxious signs of hypersexuality of many male pets. Hence, in this presentation, available methods for pet population control abroad are included.

For male dogs and cats:

It should be mentioned that **surgical methods** do not consist only of neutering, but include methods like *vasectomy* or *epididectomy*. Also easily done, and rather inexpensively too, and serving the purpose of pet population control well, they never became popular, since they do not solve other sex-related problems such as vagrancy, urine spraying, fighting, etc.

**Physicochemical methods,** like the injection of sclerosing agents into the testes, epidydimides or the vas deferens did not catch on with pet owners. But they are more than 90 percent effective, can be applied quickly, and are irreversible in their effect. Hence this is an effective method for stray and feral animals. It has been used with success to curb the stray or feral population of dogs and cats on the Galapagos Islands, which had become a menace to the world famous fauna on this unique archipelago. They are used with stray or feral tom cats in the Netherlands, which are returned to their territory after being rendered infertile, without impairing their male sex drive. This results in sterile breeding of queens, causing pseudopregnancies, which keep these females out of heat for 6 to 8 weeks. This can be repeated throughout the annual breeding season.

**Pharmacological methods** exist to control temporarily, male reproductive drive and/or sperm production. They consist of:

*Anti-androgenic hormones.* Those were originally developed to treat human sex offenders and/or to control obnoxious hypersexuality in male pets. They can be used to suppress, by a single injection, the male reproductive drive for several months, in dogs and cats, although sperm production may not be impaired.

One experience—a little anecdote from a veterinarian in England, and I think we've explained what that means:

A lady—a small person with a huge German Shepherd—comes to a veterinarian. And she gives the story that this dog is the baby of the family. The lady is a homemaker; she stays home. The husband comes home in the evening from the factory shop, and increasingly, the dog is getting hostile against the husband. And so the lovely family life becomes shattered because the husband, who never drank in his life before, started not to go home from the pub until he was loaded, so much that he didn't care about the aggressiveness of the dog. So, the Doc said: "You've come at the right moment. We just got this beautiful new drug" (It was chloramidone acetate.), "and it should solve your problems." The dog got a shot, and the lady went home, and a week later, the veterinarian got a call from the husband, who said: "What did you do to that dog?" And the vet said: "I treated it. Didn't it work?" "Oh," he said, " it worked beautifully, but you ruined my social life."

*The feeding of anti-metabolite agents* to animals. These compounds were—and still are—used for the treatment of cancer. Their usefulness for the purpose of interrupting sperm production, for up to six months after a single feeding, but *not* interfering with male sex drive, was originally observed and researched in Austria. This method has been successfully tested for coyote control by the U.S. Wildlife Service. Attempts have been afoot in Europe to use something like this to control, at least temporarily, the stray dogs. It has even been used in pets.

Where is the future? The future may be in **immunological methods.** Two such methods are available:

1. *The development of vaccines against hormones which are essential parts of reproductive functions.*

Several vaccines of this kind have been tested. They are very effective, but for a variety of reasons, only in about 85 to 90 percent of all animals. Hence, pet owners may not be willing to spend the money. For the treatment of stray and feral animals, the method is not really suitable, since after the initial vaccination, booster shots are required, which would involve trapping all vaccinated animals at regular intervals.

2. *The use of very high doses of certain hormones, or of hormone antagonists* to "overload" the system, and as a result, shut down reproductive functions.

This method of "hormonal overload" is very effective. Prototype preparations in implant systems have been tested in male

## Chemosterilants

and female dogs and complete suppression of all reproductive functions has been seen from one implant, for up to 20 months. It can only be hoped that a pharmaceutical company is willing to spend the considerable amount of time and money to develop this into a marketable product.

For female dogs and cats:

**Surgical methods:** As with the male dog or cat, there are alternative surgical methods to spay/neuter, e.g. tubal ligation. Although promoted as a "quick" method, easily performed at pounds, this method renders female cats or dogs sterile, but does not take prevent them from coming into heat. Pet owners are not interested in spending money for a procedure which does not simultaneously take the "nuisance" away of the bitch or queen still coming into heat.

**Physical methods:** These involve penetration barriers in the female's vagina. There was a time when we thought we could put a kind of cork into the vagina and prevent mating. About 10, 15 years ago, there was a lot of talk about this. There were even products on the market. It didn't work. Let's forget about it.

**Pharmacological methods:** The *anti-metabolites*, which I mentioned already, work equally well in female dogs, and, as I mentioned, presumably, in cats. They have been used in Europe, and are now used by the U.S. Wildlife Service, in coyote control. A single dosing in feed may cause (reversible) cessation of follicular development at the ovaries for up to 6 months. For stray animals, baits may be used to dispense the drug, as long as uptake by wildlife can be excluded.

*Estrus preventing steroid hormones*, like those found in the anti-conceptive pill for the human female, are available as injectable products, abroad. This has become worldwide, the most widely used pet population control method. (As I said, it didn't work in the U.S. because it was too high a dose, and dosed at the wrong time.) Treatments have to be repeated at four to six month intervals. This provides a continued source of income for veterinarians, and a health-watch for the animals, since they have to be checked routinely for the covert development of undesirable drug effects.

In the U.S., similar acting products exist only as pills (Ovaban) or drops on feed (Mibolerone). These are the only oral products available here, and they have not been very popular.

They have not caught on because they have to be given religiously every day; otherwise, it just wouldn't work. Efforts of the pharmaceutical industry to put it into dog food, to bring a medicated dog food on the market, have been consistently stalled by the Food & Drug Administration, and I think rightly so for a very simple reason: the amount of dog food eaten by people not only in low income neighborhoods but specifically in university towns where students have long ago found out that dog food and cat food is the most nutritious diet they can buy for little money, and have added it to their diet. This precludes the use of such drugs incorporated in animal food. Another point, of course, is whoever has seen a toddler and the cat eating out of the same bowl of food knows that it is virtually impossible in America to separate the intake of cat and toddler food completely, and it would be, of course, disastrous to have such a product in young children.

*Prevention of nidation* by the use of estrogens is widely used abroad, and to some extent, here. Mismating is a dilemma owners and veterinarians are not seldom confronted with. In the past and present, estrogenic hormones have been—and are—used successfully to *prevent* the nidation of developing embryos, and hence the establishment of a pregnancy in the bitch. The problems are side effects. They involve either the initiation of chronic endometritis, which may result in infertility and/or formation of a pyometra; or may result eight to twelve weeks later in a transient disturbance of blood-clotting mechanisms, which may result in the animal's death.

Presently, *anti-progestogenic steroids* are in development, which interrupt pregnancies at any stage, and which seem to be much safer.

**Abortion-inducing drugs:** Attempts to develop exclusively embryotoxic or fetotoxic compounds for use with pregnant animals have failed: products launched with fanfare in several European countries a few years ago, were unsafe for the bitch and had to be removed from the market.

As mentioned before, anti-progestogenic compounds are effective—and much safer—abortifacient agents. It is hoped that they will be marketed in the foreseeable future.

A very effective and safe class of abortion-inducing agents are prolactin-inhibiting agents, although their effectiveness is restricted to the second half of pregnancy only. A product for human use (bromocriptine) is on the world market and available for dogs, but since it causes vomiting, it is not used. Cabergoline is available in Europe. Our own studies have shown that it is efficacious and safe with bitches and queens, pets as well as stray or feral animals. Fertility after abortion is unimpaired. Abortion itself is clinically uneventful.

Let me say that probably the most attractive way nonsurgically to control reproduction in the dog, in the cat, with a single shot may be *gene transfer*. That may sound strange to you, but in the whole discussion about gene transfer and gene transfer patterns which is going on and is hotly debated, something is completely overlooked. There are two ways to transfer genes. You can transfer them in a way that they go into the germ line and then from then on this characteristic is transmitted from one generation to the next. But there is a lot of work going on with another method which we call, "somatic gene transfer." A single injection, I think, will be possible in the foreseeable future, to insert a gene at any age, any time, into an animal, or into a human being, for correction. We call this in humans, "gene therapy" and there is great appeal in the possibility of using gene therapy to correct genetic inherited disease problems. The same principle can be used to make a so-called "anti-sex" gene which would completely block reproductive functions for the lifetime of an animal. And with a single shot, without any surgery, an animal in the pound, before it is given up, would get this anti-sex gene inserted, and that would be the end of the problem of pet population control.

**VALUSEK:** Thank you, Dr. Jöchle.

# Education of Children and of the Public

**VALUSEK:** The next topic will be education—education of children and of the public. Samantha Mullen will introduce our speakers.

**MULLEN:** Before I do so, I would like to remind you of The Humane Society of the United States' formula, "L.E.S.," which has been alluded to earlier in the conference. It stands for: "Legislation, Education, Sterilization." I think many of us would be at great pains to say which of these factors deserves the most emphasis. They're all very important. Sometimes they seem equally important. I think if any element, though, would tip the balance, it would have to be education.

I would now like to introduce to you two speakers who have been eminently successful in the area of humane education: Ron Scott—director of Argus Archives, who has produced video documentation of his work and that of others on the elementary level—will speak to us about the importance of education of children; and Sheri Trainer, director of educational services at Bide-A-Wee Westhampton, Wantagh. Sheri is going to speak of her achievements in the area of humane education of adults. Ron will do his presentation first.

**RONALD SCOTT:** At Argus, we have shown humane education films for approximately 14 years. As I look about the room, I see many of the faces we see at the Argus film presentations. During the course of those 14 years, we published two books dealing with films/videos in humane education. The titles of those two books are: *Films for Humane Education* and *Animal Films for Humane Education*. Now that should have made us somewhat expert. But we gradually realized that we didn't know how the films/videos were being used, or if they were being used at all. In addition, and most important, we didn't know if the kids liked the material we had suggested for use.

So, about two and a half years ago, we decided to do something about that by going into the schools ourselves, taking films we thought were important, and seeing if they thought they were important too. And, lo and behold, most of the films/videos we thought were good (for humane education purposes), the kids liked also.

Well, what we're trying to do now is to develop a curriculum that would cover (at least initially), ten months of the school year, one subject per month always centered around some audiovisual piece. And I've brought with me today the one I have on pet overpopulation—stray cats and dogs, the subject of this conference.

The curriculum as it stands has resulted from my gathering materials from diverse sources—materials that other people have developed and used with success. The important first source was *Humane Education Resource Guide*—good stuff! This is a group of materials which Sheila Schwartz, Micki Weiss and other New York City teachers put together for use in the city schools. But this guide lacked sufficient audiovisual material and that, I think, is remedied in the second source I mention—our book, *Animal Films for Humane Education*, which includes 146 films, videos and filmstrips.

For today's meeting, I have extracted from the book five films[24] which cover the age groups pre–K through adult: *A Home is Belonging to Someone* (K–6), which we are going to see here in just a few minutes, *Friend for Life* (also K–6), *Kiss the Animals Goodbye*, *The Animals are Crying*, and *Who Cares, Anyway?* All these films/videos address the problem of overpopulation. Some organizations feel that this subject can't be broached in kindergarten. But as you will see from the filmstrip I am about to show, I believe it can be. The difficult euthanasia scenes usually associated with pet overpopulation materials have been left out of this filmstrip, eliminating the potential trauma for young children. However, the kids still get the idea.

The third source we use is NAAHE (National Association for the Advancement of Humane Education)[6a]. Here the materials are again grouped according to age levels. The subject of pet overpopulation comes under the heading of *The Consequences of Human Irresponsibility*. This is just part of the very much larger work, *People and Animals*, which was developed by Kathy Savesky when she was with NAAHE. One of the great advantages of using these materials is that the lesson plans always deal with more than one area of instruction, so that humane education is being combined with science or math, or what have you. This is very appealing to teachers, who must fill so many blocks of required instruction, that filling more than one block at a time makes life a little bit easier for them.

These are some of the sources that I've used. Time precludes my listing them all.

What have I done with these materials? Well, I've been able to use them in two school districts. The Garrison School District, up across from West Point on the Hudson River, and the Tarrytown District, a bit further south, also on the Hudson. I've shown many films, film-strips and videos, in these schools, and at the same time I've been videotaping the kids as they watch and use the materials. The teachers have been very helpful, and have come up with imaginative and innovative ideas on how to use the films and videos.

The third district that I'm trying to get into (I've had my initial meetings) is District 10 in the Bronx. Dr. Fred Goldberg is the superintendent, and I hope to meet with him later this month, along with his science coordinator, John Caterella. The idea is to run a workshop for the science teachers of the district with an emphasis on the audio-visual materials we are collating.

The goal of all this activity is the development of a curriculum which will allow school districts to begin to implement Section 809 of the New York State Education Law mandating humane education in the schools.[22] One way that we might begin this implementation is to establish these materials, including films and videos in the system of BOCES (Board of Continuing Educational Services) libraries under the subject area of humane education. Each one of these libraries (56 of them) with the exception, I think, of just a few, have developed catalogs of materials available to the teachers of their respective districts. So that would be a major step—the establishment of some category such as animal welfare, animal protection, humane education or animal rights in their catalogs.

Right now, I'd like to show you *Home is Belonging to Someone*. This filmstrip is about 15 years old, and is available from the Boulder County SPCA. It's still only $20, so that it's within the reach of most budgets.

Ronald Scott, Video

**ANNOUNCER** [filmstrip]:

*Dogs are a lot like people. Some are short and some are tall. Some live like storybook princes, with ribbons and jewels, and people to wait on them. Others work for a living, doing important jobs. Some dogs go to school and learn how to be polite and get along with people. Their masters learn to care for them. Others grow up just any old way, because nobody bothers to teach them anything. Some dogs have fancy papers, to show who their fathers and mothers and grandparents are. Others are mixtures of everything.*

*Misty was just dog. But Mike Beatty thought she was the best puppy he had ever seen. Mike hadn't asked his parents if he could have a dog, but his friend, Sally, who owned the pup said Mike could bring Misty back if his folks wouldn't let him keep her. Mike's mother didn't want a dog. "Too much trouble," she said. But his father said he thought Mike would have fun with a puppy, for awhile.*

*So Misty had a home, not a very good home, because Mike didn't know much about dogs, and nobody else cared about Misty or taught her how a dog should act. But Misty loved Mike, and Mike played with her and fed her when he remembered, and gave her fresh water when he remembered too. And so Misty grew up. Nobody knew she'd be such a big dog or that she'd eat so much. This worried Mike's mother, but Misty kept out of her way.*

*But then something happened: Mike's father's job changed, and the family had to move to another town. They were going to live in an apartment. There would be no room for Misty. Mike tried to get some of his friends to take Misty, but nobody wanted a grown-up dog. It was almost the day they had to leave. Mike's mother said it was still summer, with lots of people going on picnics. Why not take Misty out to some nice place where there were lots of people and just leave her there? Somebody surely would find her and feel sorry for her and give her a home. In the end, that's what they did.*

*Misty ran as hard as she could to catch the car,*

*until it was out of sight. Her lungs felt as if they'd burst. For a long time, she sat beside the road, hoping Mike's father would come back. Several times, a passing car almost hit her. She was terribly thirsty. Below the road, she found a little trickle of water, and after she had had a drink, she sniffed around for just one familiar smell, but there was nothing.*

*Misty knew she was lost. She was so miserable that she put her nose in the air and howled. But crying never solves anything. So she followed the stream until she came to a place where someone camped. In the trash, she found a piece of moldy bread and a burnt wiener. Not much of a meal, but Misty was too scared and lonely to be very hungry anyway. All night long, she walked beside the little stream. She missed Mike and her old rug at home. Somewhere, she stepped on a piece of broken glass. Most of the day she rested, nursing her bleeding foot. The going was slow after that. She was hungry, and she got tired quickly.*

*The next day, something good happened: Misty came to a place she remembered. She had been here before with Mike and his friends. Home could not be too far away. Misty hurried now on three legs, stopping only to pause for familiar smells. It was Mike's schoolyard, where Misty and Mike had played together, but nobody was there now. Misty trotted past. And here was home!*

*Misty even used her sore paw to run the last little way, but no one was there to welcome her. Mike's bike was gone, and so was the family car. Misty lay down on the porch to wait. She waited until she was so hungry she couldn't wait any longer. Then she started out to find food. She went to the edge of town where there were more people, sniffing at every trash can. At last, she found a garbage can that smelled of meat, but when she tried to look inside, it tipped over, and the angry owner chased Misty away. Later, she found a house where the family cat had not eaten all her supper, and Misty stole what was left. Then she went back home to wait some more. Her sore foot felt hot and throbbed painfully. For several days, she lived this way, eat-*

ing whatever she could find and sleeping on the porch. She was terribly lonely, for **one thing every dog has to have is someone to love.**

One day, a strange truck came, and some strange people began unloading things. Misty was glad to see them. But they weren't glad to see Misty. The woman yelled at her and chased her away. Now, she had no place to stay, so she just wandered up and down the street, wherever hunger drove her. Sometimes, when she tried to steal food, people threw rocks at her. She grew thin and her coat was rough and tangled. Misty had become a tramp.

As fall came, days were often chilly, and nights grew longer. Few people came to the picnic grounds. Winter was coming. One day, it was much colder. Misty huddled in a little hollow protected from the icy wind. Tonight, it would snow.

On the highway, a car swerved and stopped. Misty went down to see who could be stopping on a day like this. The car had engine trouble, and the woman and little girl couldn't fix it. Misty drew closer. She liked people, and this looked like someone she could trust. Just as a kind man stopped to help, the girl saw Misty:

"Mother, look at that poor dog. I wonder who she belongs to."

"My goodness, she looks half-starved. Be careful, dear, you never know how a strange dog may act."

Misty wagged her tail and whined softly. It had been a long time since she had heard anyone speak kindly.

"That dog's been around here for weeks. Somebody must have dumped her. Seems friendly enough."

"Poor dog, I wish we had some food. Mother, can't we take her home?"

"You already have a dog."

"But can't we just take her and feed her, and then we could take her to the Humane Society. Please,

## Education of Children

"Mother?"

*Misty was so thankful to get into the warm car that she didn't care where they took her. At the girl's house, Misty ate the first real meal she'd had in months. She wished she could stay with these kind people forever, but they put her in the car again. They drove out toward the edge of town. The girl kept talking about her scout troop's visit to the humane society, and how the people there cared for animals with no homes. At the shelter, the girl told how and where they had found Misty.*

*Misty was sorry to see the girl and her mother leave, but the man spoke kindly. At first, Misty was afraid. She didn't understand the strange smells, and there were so many other dogs, some of them barking and whining. She didn't like being shut up in a cage, but the cage was warm and clean, and there was a short run where she could see outdoors. The people seemed friendly, and she had two good meals a day, and plenty of fresh water. It certainly beat living like a tramp.*

*The kind man looked her all over carefully. He bathed the sore foot, which had never healed properly, and put something on it to help it get well.*

*Sometimes, a girl or a boy would come and take Misty for a walk outdoors. These were the best times. But, oh, how she longed for someone to belong to, someone she could love and live with!*

*Very interesting things happened at the Humane Society shelter. There was an ambulance that went out to help sick or injured animals. There was man who went out to investigate reports that an animal was not being treated kindly. Besides dogs, there were other animals. There was a whole room full of cats and kittens who, like Misty, needed a home. And there were puppies. Misty heard someone say, "Only one puppy in seven ever finds a home, and it would be better if they had never been born." The humane officer explained that a simple operation called spaying prevents female animals from ever having babies. Almost every female dog and cat should be spayed, so there won't be puppies and kit-*

*tens without homes.*

*Sometimes, there were raccoons and other unusual animals at the shelter.*

*Every day, visitors came looking for a pet to adopt. How Misty wished someone would adopt her, but there were always many more dogs and cats than there were folks wanting them. The boy with the brown hair and freckles seemed just like all the others, until he knelt down and said something very softly to Misty: "You're the one." Misty was so happy she thought she'd burst—a real, live boy who was to be her very own!*

*Before Misty could go home with Doug and his parents, his father had to sign some papers. They had to promise to provide kind and loving care, to give Misty plenty of good food, fresh water and a comfortable place to stay, had to take her to a veterinarian to keep her well, and to obey all the laws about dogs. They had to promise they'd never, never go away and leave Misty, as her first family had done, never permit her to be used for any laboratory experiments. At last, Misty knew she was going to have a real home, a family to love and belong to.*

*All the things the Humane Society does for animals that need help take lots of money. The money comes from gifts given by everyone who cares what happens to hurt or homeless animals. If ever you find a stray animal or know of any animal in trouble, remember, tell the Humane Society about it. The Humane Society wants to help all animals in trouble.*

**SCOTT:** The filmstrip is a little bit sexist in that they only mention spaying, and not a word about neutering. It's still usable.

We have a whole other side of Argus too, and I'd like to urge other people to join in. That is, in videotaping the various activities that we are all involved in.

Some of the things that *I've* done, for example, since I think we had an *Argus Archive News* for the first time at our last film

## Education of Children

showing, is that I got the bucking strap at a rodeo outside of Albany. I think it's the first time that I've seen it. I have pictures of the horse with the bucking strap on and as the bucking strap was released and the animal became pacific. Last week, on Labor Day, I was at Hegins to video the pigeon shoot. It's not pleasant to do it, but I think that we need this footage. What I'm saying is, I urge other people to do it around the country. We're going to try to put it all with Focus on Animals, a new organization started by Esther Mechler. With this kind of footage housed in one place, I think that will be very valuable for people who wish to make new videos, films, etc., And again, I urge everybody to do that.

**MULLEN:** Thank you, Ron. We'll hold the questions until after Sheri Trainer's presentation is made.

## Sheri Trainer

**TRAINER:** Thank you very much. Ron, that was an excellent presentation. I have known Ron for a long time, and have admired the work at Argus, I guess, for ten, fourteen years literally. And I'm very happy to hear him speak about a program that's extraordinarily important because teachers *do* have an awful lot of power, and to work very, very closely with them gives us a lot more leverage when we really want to speak to the general population.

All during the conference, I guess with the majority of the speakers, everybody had to refer to humane education. But let's talk a little bit about what education really is, and I will give you some ideas about how I'm tackling my job at Bide-A-Wee.[14]

First of all, when we talk about the parameters of education, we are discussing prescribed programs such as Ron has been talking about. Of course, we're also talking about the mass media, the TV and the radio. It's very, very important. I really hope that we pursue it more in the future than we have been doing in the past. We aren't talking about publications; we are talking about educational forums such as *Today*.

Of course, public demonstrations and public actions also have a lot to talk about, when we want to really get our message to the public.

But, coming a little bit closer to home, one has got to say: "What is the message that we really want the public to know?" We have an enormous job. We are called upon very often to talk about public health, such as dog bites and, in particular, the pit bull issue, which everybody in this room has heard *ad infinitum*, and I don't want to go into it in any more detail here. We have to be knowledgeable about zoonoses. Of course, rabies is only one of the topics that we hear about most often, but it's certainly not the only one. We have to know about strays, straying animals, feral animals, and we have to be involved with animal capture, and that is a very, very difficult thing to be involved with, especially when many organizations do not have the best reputation in their communities and they come across very irate public citizens. So it's a very, very tough job.

In addition to that, we have to know about pet care, whether the animals are going out of the shelter on adoptions, or whether we're just being asked general questions from the public. I must say that even though veterinarians are out there

## Education of the Public

in droves, the humane society, the shelter, is the first place that people go to when they have questions. So we have to be knowledgeable about pet health and about animal behavior of various kinds of breeds—in particular of dogs and cats.

In many organizations, for instance, the ASPCA, there are other types of animals that are being adopted all the time; they have a tremendous exotic animal program. Of course, Bide-A-Wee has the advantage with me because I do specialize in rodentology, and I love rats. And even though we don't adopt them, they are the companion animal of the future and, unfortunately, they've been terribly maligned.

We also have to know a lot about animal nutrition, and more and more, we have to be knowledgeable about anti-cruelty. It is an absolute must. And let me just mention too that, as the field has grown, we are becoming more mature in our attitudes towards other issues: the human health issue has become a very, very important part of many humane education programs—how animals are raised for food; how animals are used in science. It has to become a part of what we talk about, because we're talking about an ethic and the ethic is the reverence for life principle, and that's going to permeate every part of the image that we want to portray to the public.

Environmental issues, ecological issues—we have to talk about them, because we don't want to isolate any part of the community. And in order for people to really understand what the relationship is to the "sacred cow" of the dog and cat, we have to look at a general perspective.

The first area I'd like to address is professional training, and out of all of the professions that I think we must concentrate on, it is our own. There is no place that I know, outside of one or two schools, that prepares people on a pre-service level to work on shelters or shelter management. It doesn't exist. That is ridiculous. Even if I look at the animal care programs in high schools, of which there are a few, not many, it doesn't exist there either. We've got to change that. We've got to get two year programs, one-year programs, either in animal health technician specialization or the public health specialization, and get programs designed specifically for people who are interested in going into public service, which is animal control. I get questions all the time: "Where can I go to get some training?" I don't know what to tell them, other than the fact that the only thing that might exist if they cannot get into a veterinary school, in most cases are animal technician programs, and it's just not enough. We really have to concentrate on pre-service. That is not to say that in-service training and apprenticeship training is not important.

In every single profession that I know of, there are training programs, once somebody has already been prepared in that field. I was in one of the major teaching hospitals several months ago, and you know how the interns come around on their rounds, and I saw them coming, and I thought: "Here come the apprentices." They're apprentices, just like we have to develop an apprenticeship program in our own field that really makes it work for us; have some kind of standardization. HSUS is doing a nice job with it. I know there is a training center up in Delhi; I've not really been involved, but it's important that we really standardize it.

The other area of professional training that's very, very important is teacher training, and I know we talk about it all the time. I've got two programs that are going to be given, one in the fall and one in mid-winter, on Long Island. One is going to be through Molloy College, and the other one is going to be through the BOCES program, Western Suffolk County Teachers Association.

I want to say something about college programs. I work with colleges to the hilt. The colleges love me because I make money for them. People want to know about these issues, and they will pay to learn about these issues, and the reason why I'm working through colleges is:

**1.** they will attract people that I cannot, working for a humane society, and;

**2.** this will give them an impetus to actually make the animal welfare and rights issues part of their curriculum when they see the need for it.

So please explore your colleges, because they really want you, especially if you pick your programs correctly.

There are additional areas in professions that we really have to hit. At the Bide-A-Wee Home Association, we work with psychologists, we work with social workers, we work with recreational specialists. I'll talk a little bit more about this later.

The area of animal-facilitated therapy has been both a blessing and—I hate to say it—a curse in certain respects, because it does take a lot of time away from really, really needed programs in humane education. However, make animal-oriented therapy programs work for you. Insist that any time your organization goes to do a program in a facility, that you talk to the staff, and when you do, always make it part of your program to talk about the bonds that *don't* work, the cruelty. You will be amazed at how interested people really are and really don't know outside of a common sense approach that they may or may not have, that research really has been done in this field. I used that Kellert study.[25] It is wonderful.

We have also done some programs at Molloy College, which is a Catholic college out on Long Island. When we did those programs on animal-facilitated therapy for all of these specialists, they wouldn't let us leave the building until we planned the next term for them.

So It really does work, and people in other professions really do want to hear you, but you've got to make it palatable for them.

The next area that's very, very important, is the community education program. You *must* you get to know your community. We have three facilities at Bide-A-Wee: one in New York City, one in Wantagh, which is around the Jones Beach area, and the other one is on the south shore, all the way out in Westhampton. I would never do the same program for the community in Wantagh that I would do in Westhampton or that I would take to New York City. I am dealing with a completely different constituency. These people have different lifestyles, they have different perceptions, they have different levels of awareness. That does mean that eventually I will not bring them all up to the same level, but I will have different approaches. You will only be able to have successful programs when you know what your community wants.

We did an adult education program,[26] and I have had some people from this audience participate. This field really has experts. Take advantage of it. I am not an expert in many fields, even though I try to be well read. Dr. Rackow and Dr. Mark Lerman were there, and I intend to continue to ask them to participate, because I can't do the job that they can do. It adds a lot to my program to get these experts working with me. But what I did out in Wantagh was—before I started my adult education program—I sent out questionnaires to the community, and I asked them what they wanted to know.

Based on that, I did my program. I hit every area that I wanted to, but I did it in a different sequence. After I was through with my program, I sent out little critiques to see if what I provided them was satisfactory. If it was not, of course, I changed things. Incidentally, I always change things anyhow, and I always learn something from the first time.

Other related groups—this is part of your community—please hit them: Rotary Clubs, Kiwanis Clubs, glee clubs. Glee clubs really do have very, very responsible citizens who are extraordinarily concerned with animal control. Get their support. They are important to you. They do fabulous things, and they will work with you in getting good programs together. They will also support your efforts financially.

Another area that's very, very important—and I must be honest with you—I haven't done it yet, but I will be doing it. Get into private industry. Much of private industry has what they call employee enrichment programs. This is your community. They'd love to hear from you. Do a little dog obedience, animal behavior, get into human diet and then get into the factory farming issue. It really works.

Continuing adult education programs—You can offer them through the high schools. It really doesn't matter. I have offered it, as I said, directly from Bide-A-Wee to the community. People are always anxious to learn, so make it available to them. You don't always have to do these types of programs in your facility. As a matter of fact, I encourage you not to do it in your facility all the time, because if you're reaching out to the community, you've got to go to the community. It will not always come to you.

This morning, flea markets were mentioned, and I just want to say a few words about that: Flea markets, street fairs, malls, yes, they are very important, and we are involved with those kinds of activities, and what we do, of course, is distribute certain types of publications. We do have volunteers trained to speak to the public, and we do try to solicit people for our mailing lists, because we have a lot of really good programs at Bide-A-Wee.

As to school programs, my emphasis is on teacher training. Teachers, as I said, are an extraordinarily powerful group, and in order for teachers to get differentials, they have to take a certain amount of in-service credit. And unfortunately, I had looked through an offering of the type of courses that teachers are taking, and it's amazing to me how this profession has watered

## Education of the Public

down a lot of the types of materials that should be a lot more meaningful to them. The people that you will attract to these school programs I will tell you are going to be with you for life. They're really dedicated. I give out assignments on my in-service training. I expect people that work with me to work hard, and I get beautiful projects back.

I *do* do school programs. I don't do them as my only activity. I supplement school programs. Which means, if teachers are teaching any of the issues as part of their curriculum. I make all the materials available to them. Videotapes, of course, I screen first. I supplement what teachers are doing. "One-shot" deals do not work. They do not. You're wasting your time; you're spinning your wheels. Supplement what they're learning. Talk to your teachers. Show them as many materials as you can. There are a lot of materials out there, and different teachers at different grade levels in different parts of your community need different things. One thing cannot be prescribed for everybody. It's just not going to work.

You have to analyze your own staffing in your own facility, and your own budgets. I am very, very fortunate to be at Bide-A-Wee, and to be working with fabulous people in my department. Ursula Goetz, my executive director, is supportive of everything I do. It really makes it a lot easier when that happens.

What we also do in my department—and again, I am able to do it because I have the staff—is offer additional programs. For instance, Debbie Feliziani does many things. One is organizing a problem dog and cat clinic. She will also be getting involved with consultations after people adopt their animals, try to steer them in the right direction.

Seymour Sub is our legal consultant. We are trying to set up training programs for the police, for the district attorney, and for the judiciary regarding anti-cruelty, a special program.

Your community's going to love you, because you're not charging anything for those, because you charge them for other things, but not that. Hook them first.

We also have a wonderful individual who does bereavement counseling. She has a pastoral care background. She is also doing a beautiful reverence for life program. Again, we do have a pet memorial park, and it does become something of a necessity in our case. Not everybody can do it. In our fortunate circumstances, we can.

The last area that I just want to mention very briefly is animal-facilitated therapy. Barbara Cassidy had mentioned this morning that it is not the responsibility of humane education departments to perform this task, and I do agree with her. As a matter of fact, the first time that I got involved in pet therapy, I had gone to a conference, I think it was about two years ago, and I met Phil Arkow there. He's been involved with pet therapy an awfully long time, and in my frustration, (because I had just come to Bide-A-Wee and I had to take over this program that was really already quite extended) I said to Phil: "Gee, here I am in a situation where I have to take animals to geriatric facilities, mental health facilities, and even though I know something about the problems that I'm dealing with, I don't know enough about these people to really understand what effect my intervention is going to have on them. Why am I doing this?" And he said: "People in our field have to be the ones to encourage change."

It really hit a point with me, and when I came back to Bide-A-Wee and I got Debbie working with me, I said to her, "What I would like to see happen in my department is that we sit with the people who are most responsible for it, since it is a human service: the social workers, the psychologists, and the recreational people."

What we've been able to do with Bide-A-Wee is get C. W. Post and their internship program in counseling involved with this. That's where it should be, and what humane societies should be doing is consulting for this kind of thing, helping them with temperament testing on animals for this program, or if there's a need for a residential program, to really delineate what the needs would be. But actually to be involved only in a visitation program does take an awful lot of time, and you really do have to look at your own needs and what your needs are in the field, because you have to prioritize. In the long run, I do see social services taking this field under their wing. That's really where it belongs.

One other thing I want to say in closing, and this really doesn't have too much to do directly with any of my precepts on humane education, but one thing that I know that I practice as an ethic—and I hope we all do here—is that we have to cooperate with one another. We have some really good people. We have some fabulous people. Don't hold grudges if any existed in the past. I can tell you this field is growing. Open your doors, talk to one another, network.

## Education of the Public

**MULLEN:** There is time to ask a few questions of our very dynamic speakers.

**MAX SCHNAPP:** We have a humane education law in New York State.[22] But one of the main things this law confronts us with, is how do we penetrate that iron curtain of ignorance in the New York City school system, the Board of Education. If any place needs a minimum of humane education, it's New York City and those children. I see it day in, day out. Therefore, my question is: Have we made any progress in winning over or in getting a foothold within the Board of Education, to make it possible for us to teach those children a little humane education?

**TRAINER:** I've been working very closely with Sheila Schwartz. She is completing her doctorate in humane education at Columbia, one of the few doctoral candidates in this field, which will add an awful lot of credibility to where we want to go. Sheila's been working very, very diligently with the Humane Education Committee, lobbying within the City to get funds appropriated for the schools. We talk about materials such as this guide, we talk about audio-visual equipment, of course.

Let me point out one thing to you. Teachers in the New York City system have many types of pressures in addition to the reading and writing and the standards that they have to satisfy for the State: drug education, suicide, sex education, every group wants its priority, and, unfortunately, humane education is just another one of them. We do need extra funding, and there will be a resolution in the City Council sometime this year to appropriate more funds for this activity. It cannot be done without funds. The teachers are interested. I know they're interested, because it's a winner. You can't deny it.

**MULLEN:** It's self–evident that, to change the way people behave towards their animals, we must first work towards changing the way they think about them. With dynamic and creative educators like the two we've just heard from, I believe we stand an excellent chance of making some real progress in that area. Thank you, Ron Scott and Sheri Trainer.

# Attitudes Toward Neutering and Euthanasia

**MULLEN:** We're now going to hear from Dr. David Samuels. Dr. Samuels participates in several community spay/neuter programs, including the low-cost program offered by the Ulster County SPCA. He is going to treat the topic, "Medical Myths About Neutering and Euthanasia."

The following speaker will be Dr. Murry Cohen, a practicing psychiatrist in New York City, who is also secretary-treasurer of the Medical Research Modernization Committee, a group concerned with the welfare of experimental animals.

Afterwards, Ingrid Newkirk, of People for the Ethical Treatment of Animals, will speak again.

Following Ingrid is another famous proponent of animal rights, Dr. Tom Regan, professor of philosophy at North Carolina State University, in Raleigh.

Then, Elinor Molbegott, of the ASPCA, will review legislation affecting animals. Afterwards, Assemblyman Maurice Hinchey will join us.

We'll begin by hearing from Dr. Samuels.

**DR. DAVID SAMUELS:** I have two veterinary practices: one in Kingston, and one in New Paltz, New York, and I have been practicing for ten years. I graduated from the University of Pennsylvania. I'm going to open this part of the discussion by talking about some false beliefs that pet owners express to me when they come in. I also want to mention a few personal experiences that I feel will better illustrate some of the problems that pets face because of their owners being uninformed. And, lastly, I want to touch upon euthanasia and something that happened to me recently in dealing with that.

I'm going to speak to you as a guy in the trenches out there dealing with the public every day. I am not sitting in an ivory tower someplace, and making up a lot of these things. These are issues that I deal with every day—day in, day out. Something happened last week which happened to someone that wouldn't happen to anyone sitting in an office someplace, who doesn't come in contact with people and their pets on a day-to-day basis.

I'm going to start out with talking about people coming and saying, "Well, I'm not going to get my dog or cat spayed because it's going to get fat." And the truth of the matter is that I see many pets out there that are grossly overweight that aren't neutered or spayed either. And I think that if people would just take it into account that yes, maybe their pet's metabolism may slow down a little bit from the sterilization surgery, but if they would monitor the food intake and the activity level a little bit more, most of these pets would not get grossly overweight. That's a big thing that I see constantly. And we try to educate them, day in and day out, as to as to why they should go ahead and get their pet neutered or spayed.

The next thing that I usually hear about is (usually from the guys that come in that are the big hunters and what-not): "If I don't let her have a litter, she's not going to make a good pet for me." That's totally unfounded, and that's like saying if a woman doesn't bear a child, she's not going to be a good woman. It just doesn't make any sense. As a matter of fact, a lot of these female pets that do bear offspring, I find that a lot of them go through a lot of behavioral changes: They become moody, they can become overly protective. I have had instances where they have bitten family members who have gone in and tried to handle the puppies. Then, when the puppies are taken away, they definitely go through withdrawal symptoms. And I do think it may actually affect them afterwards.

There are a lot of times when people will come in and say: "My cat is a week pregnant or two weeks pregnant, so obviously, we can't do anything." We try to educate them that that's not the case. And, unfortunately, all too many times you're spaying these cats—or dogs—almost to term. If we could only educate these people to understand a little bit better what they're doing and what they're putting these animals through, it would help us a lot.

## Attitudes Toward Neutering and Euthanasia

People will come in and say: "Well, the surgery is very painful, isn't it?" And most people out there, I don't think are aware that veterinary medicine is in the 20th century, and we don't put a cat in a boot any more and neuter him that way. Surgical procedures in modern veterinary practice are always performed under anesthesia. The pet being spayed or neutered does not feel any pain. If there is any kind of a counterindication to surgery, we always correct it first, obviously. Your practice wouldn't grow very much, if every pet you put under anesthesia didn't do well. Medical emergencies really don't occur any more frequently than they do in human operating rooms, and that's something I've been trying to get across to people. I personally let them watch the surgery if they want to. It doesn't bother me at all. Samantha has assisted me many times.

This brings me to the medical aspects of why or why not to spay, and there are definite medical reasons to get your pet spayed, and I'm sure most of you know this. The incidence of mammary disease—mammary cancer—definitely goes down if a dog or a cat is spayed before its first heat. Even if they experience the first heat or second heat and then they are spayed, the incidence of mammary cancer does not change very much as the pet ages. They're just as likely to get mammary disease as a pet that never was spayed. In other words, if the pet was spayed after the second or third heat, there's not going to be any change later on. It's the same as if that cat was never spayed. But if you spay them before their first heat, the incidence of those types of problems drops way down, and most people don't understand that.

I had an incident happen to me last week in my practice where a lady brought in her nine-year-old poodle. The dog was vomiting, it had a vaginal discharge, and was going downhill. After doing the blood work and going by the clinical signs and abdominal X–rays, we told the owner that the dog had a uterine infection technically called "pyometra," and that we would have to do immediate surgery to save her dog's life. And she related to me at that point that the reason why she didn't have her pet spayed earlier was because she had heard of somebody else whose pet was killed by anesthesia. Because of that, she was never going to let her pet undergo anesthesia. And at this point, I told her that her pet was in a lot worse shape now to undergo anesthesia than it would have been when it was six months old, and we had spayed it and removed its uterus at that point.

And, unfortunately, we did the surgery, but the toxins that the bacteria had produced had put the dog into kidney failure, and even though the surgery was successful, the dog died from the toxins of the pyometra. The owner did not take that very well. She blames herself very severely.

I definitely see more mammary tumors in pets that have not been spayed. The unfortunate thing about that is that, in some of these cats that aren't spayed that develop mammary tumors, a very, very high percentage of the tumors are malignant. To do a radical mastectomy on a pet is something that you really don't want to see, when you lay them open from stem to stern and you end up ligating about fifteen arteries and just peeling everything out from here to here. It's very painful for the pet. It's a gruesome surgery; it's something that we don't like to do. And even after you get done doing that, you end up having to tell the people that once everything's healed, we want to go back in and spay it to try to prevent any further recurrence, because when they go into season, and their estrogen spikes, sometimes that's what stimulates more mammary problems.

As far as males go, neutering a male really does keep him closer to home and keeps him from wandering and seeking out females in heat or other male dogs to fight with. And I'm sure that all of you are aware of that, but most people out there are not. The ordinary person out there is not aware of it, and there are a lot of guys that will walk in with their wives, and they'll say: "Well, you're not going to do that to my dog. There's no way you can do that to my dog." It's tough, but it really does help to explain the benefits to them of having the animal neutered.

In the male cat, it definitely does prevent them from spraying, although a certain percentage will still spray if they're threatened or other males are around. But it definitely reduces that. It definitely helps in overpopulation with the cats. And, incidentally, neutering a male cat does not bring on FUS. I don't know how many of you are aware of that, but it's a syndrome in male cats where they will plug up and not be able to urinate. It has been disproved that neutering them when they are a mature size, is a cause of FUS in the cat. I mean, obviously, if you neutered them very young and if their growth is stunted, then yes, there might be a physical component there. But if they're neutered when they are a mature size, that's really not a proven cause of FUS in the cat. It definitely reduces prostatic problems in a dog, testicular tumors, and sometimes some of the more carcinogenic tumors in male dogs.

The last subject that I want to touch upon a little bit is euthanasia. A couple of weeks ago, I was asked by Samantha to go to the local shelter and euthanize some animals there. And when I got there, I was told there were approximately forty to be euthanized. Later that day, in the evening, I was supposed to go play racquetball and go out. As I walked in the shelter, it was raining that day, it was miserable. After about the first ten animals, I started feeling awful. And the mood of the shelter people and everybody just started going downhill, and by the time we reached twenty, nobody was talking too much. And by the time we reached thirty, we were all looking out the window.

That was the first time that I could really identify with what the shelter personnel go through in euthanizing these pets. I didn't know them, the pets. When I walked in there, the shelter personnel knew a lot of animals by name, they had become bonded to them, they were friends with them. And so I can't even imagine how they felt, but I know when I walked out of that place that day: I felt like hell, and I didn't play racquetball that day, I just went home and kind of sat in a chair, and that was all I did. I thought that I'd kind of developed a little bit of a barrier between myself and them, but I guess I really hadn't, and I think if the general public could see something like that, it'd open their eyes up real fast.

I don't know why there's animosity between veterinarians and the shelters. I think we're afraid that you're going to step on our toes, and I guess there are a lot of us out here that are kind of struggling to survive right now because of other problems. But I think that if we spayed, like we talked about, from morning to night, we still wouldn't make a dent in it. I would hope that some of the younger veterinarians like myself will have more of an open attitude about it and try to work along with you people instead of saying: "We don't want you doing heartworm tests, we don't want you performing euthanasia, because the next thing you're going to do is, giving this and doing that, and then we're even more in trouble." I think a lot of the people, a lot of my fellow veterinarians, would accept something that was laid down well: some sort of a national guideline with a training program that the veterinarians feel comfortable with, the shelter personnel could feel comfortable with, sanctioned by the higher-ups, within guidelines so that we all felt we weren't going to be threatened, so that we could work together.

**MULLEN:** Thank you. I hope that, more and more, we'll able to cooperate in eradicating those problems.

## Attitudes Toward Neutering and Euthanasia

**MULLEN:** Our next speaker will be Dr. Murry Cohen, who will discuss how psychological and religious factors affect attitudes toward neutering and euthanasia.

**DR. MURRY COHEN:** Thank you. I'd like to offer a "Benvenuto" to our Italian colleagues here from the Animal Advocacy Movement. We embrace you, and I hope that you learn from us, as we can learn from you.

One word about a prior presentation, the presentation of Dr. Lieberman. I'm not a veterinarian, so it is difficult for me to comment technically on what he said about early spaying and neutering. But it's not a surprise to me that the veterinary association resists it. As my medical brethren here, such as Herb Rackow and Neal Barnard, can attest, organized medicine resists many new and innovative things, and it takes many years for these things to trickle down and become incorporated into mainstream efforts. It was true of the electrocardiogram, by the way, which is a device that, as you all know, is pretty routinely used nowadays. As a matter of fact, it's invaluable. When it was first introduced, it was rejected by the AMA as being superfluous and unnecessary, and actually an insult to cardiologists. And the same thing happened recently with the NMR, the nuclear magnetic scanning procedure which took years and years to gain acceptance.

I've tried to orient my talk from the animal rights point of view, but I'll leave the explicit development of that point of view to my colleagues, Ingrid and Tom. I'm here to speak about psychological and religious influences on attitudes toward neutering and euthanasia. I assume that the reason why psychological and religious influences have been combined into one subject is that just like the basis of religion is faith, so the basis of much animal-based psychology experimentation is also based on faith. It certainly is not based on reason and logic, considering the ludicrous nature of much of the animal psychology experimentation that gets funded. It would take an almost religious devotion to the experimenters to give any credence to their psychological experiments using animals.

As a group, dog-lovers are certainly concerned about the problem of companion animal overpopulation. A *Dog Fancy* readership poll published a year ago revealed that 43 percent of respondents were primarily so concerned, compared with

46 percent who were primarily concerned about individual cruelty or abuse, and 6 percent about laboratory research. A total of 77 percent of the respondents belonged to an animal advocacy organization: 39 percent to HSUS (and this was a year to two years ago); 29 percent to the International Fund for Animal Welfare, 28 percent to the ASPCA, 25 percent to the People for the Ethical Treatment of Animals, and 21 percent to the Animal Protection Institute. Why is it that, despite these memberships in organizations, all of which promote neutering and humane euthanasia, despite expressed concern on the part of people about companion animals, and despite the pleadings of knowledgeable animal welfare advocates, companion animal population remains a severe problem?

The view of people experienced in working in shelters is well expressed by David Kaye, education director of the no-kill shelter called Treehouse Animal Foundation in Chicago, who states:

> *From my professional perspective, there can't be enough emphasis on the tragedy of pet overpopulation. Several of the major concerns of people would be resolved if the pet population explosion were curbed. Euthanasia in shelters, illegal sales of shelter animals and much frivolous animal-based product testing and research would be significantly diminished if it weren't for the current 12-to-1 ratio of homeless animals to available homes.*

And traditions certainly do foster humane treatment of animals:

Catholicism gave us the wonderful St. Francis of Assisi, whose "Canticle of the Creatures" exalts God through the world of creatures, in whom St. Francis recognizes identity and feeling, and whom he addresses as brother and sister. Cardinal Gibbons wrote of his "abhorrence of unnecessary cruelty to dumb animals."

Protestantism produced Albert Schweitzer, whose prayer we read together last night and who blessed us with his "reverence for life" philosophy.

Judaism has a long heritage of compassion for animals and concern towards *tsa'ar ba'alei hayyim* (suffering of living creatures). *Proverbs 12:10* says: "A righteous person knows the soul of his beast." And *Ecclesiastes 3:19* says: "For in respect of the fate of man and the fate of beast, they have both one and the same fate. As the one dies, so dies the other, and both have the same life breath. Man has no superiority over beast."

Hinduism produced Gandhi, who advocated kindness and compassion toward all animals and regarded cows as being imbued with maternal traits, which should cause us to seek nurturance from them.

Buddhism teaches that all creatures love life, and that life is sacred.

Mormon prophets believe that animals have spirits and will be resurrected, that people will be judged on how they treat animals, and that animals will communicate with humans in the heavens and enjoy "eternal felicity." Wouldn't it be great if the Mormons are correct?

Yet, Catholics regard animals as being unworthy of significant concern because they have no souls. The Holy See is notoriously silent on the subject, and mainstream Catholicism has effectively renounced the Franciscan tradition.

Yet, despite the teachings of Rabbi Eliezar Hakapard that a person is not permitted to buy cattle, beast or bird unless he can provide adequate food, the Jewish tradition of *Chalaka* forbids the castration of male animals, inhumane slaughter continues and pro-Messianic behavior requires eating of chickens.

Yet, animals are terribly treated in Hindu India.

Yet, the state of Israel—which in the biblical tradition respects and preserves wildlife to the extent that it's against the law in Israel to pick a wild flower—is unresponsive to the need to protect companion and so-called work animals, does not emphasize humane education in schools, does not adequately support shelters, does not allow the entrance into the country of a duty-free animal ambulance and handles the cat overpopulation problem by poisoning feral cats with strychnine, which produces a slow and agonizing death, and ironically, leads to the proliferation of disease-carrying rats. Despite Israel's disdain for blood sports such as hunting and bullfighting, an American animal protection group that Ingrid mentioned yesterday, CHAI, is needed to influence Israeli policy to reflect more humane approaches to non-wild animals.[9]

Yet, with the exception of such clergymen as the Rev. James Parks Morton, who officiates over the Festival of the Animals and the Blessing of the Animals every October at St. John the Divine, on the feast day of St. Francis, Protestant theologians have not addressed themselves to the crisis of animal abuse in general [27] and companion animal overpopulation in particular.

With all of these contradictions, and considering that the human psyche is full of contradictions, it's no small wonder that conflicting attitudes pervade the field of animal care. In fact, because of the complex relationship between human and non-human animals—mystical, utilitarian, psychodynamic, economic and others—the contradictions are probably even greater than usual.

Probably underlying most, if not all, of the resistance—both religious and psychological to neuter and euthanize—is the degree to which we identify with companion animals, who over the millennia have been especially bred to conform to our wishes, satisfy our needs and provide us with unique non-human life forms to relate to. Small wonder that identifying with them is so easy. By identifying, I mean taking their nature into our psyches, so that they actually become a part of us. This is a very special and potent bond. We see it very clearly in children. Much of this identification is desirable, for it leads to empathic communication, respect, the need to nurture and protect, a sensitivity to the needs and interests of the animals, and a more general awareness of animal abuse outside of the companion animal area, such as factory farming, rampant animal experimentation, hunting, trapping, animal mutilation by tearing their coats off their backs, and disgraceful animal "entertainment" spectacles.

However, there is a dark side to identification, and it is this aspect which underlies many of our less rational and adaptive attitudes toward euthanasia and neutering. With respect to euthanasia, we are caught on the horns of a dilemma, for regardless of how we become interested in animal protection, be it reverence for life, animals rights, animal welfare or whatever, we come to the belief that the lives of animals are precious, and yet we must kill them. And in order to kill them, we must partially sever the bond of identification we have with companion animals by remembering that, although they may be equal in some moral sense, they are not identical. Their plight is quite different, different because we created them, different because we created their excesses by removing natural limiting population factors, different because society at large looks at unwanted animals differently than unwanted people, and different because they are, so far as we know, incapable of conceptually anticipating the state of being dead. And kill them we must, for the alternatives already mentioned at this conference are unacceptable.

But our moral and religious backgrounds teach us that killing is wrong. The word euthanasia can mean good death, but it also means "well" death, as the prefix *eu-* can mean well, as in *euthyroid*. Our logical facilities are strained by the idea of a "well death." It sounds like a contradiction. But death within the context of the alternatives of a life of pain and misery can represent the option of wellness, even if the means to that death is killing. As Ingrid said yesterday, life is not always the desired state, and death is preferable, if it is properly utilized to address an abnormal situation for which satisfactory alternatives simply do not exist.

Many might contend that there *are* satisfactory alternatives. From everything I have read and heard, this is not the case, so why do people believe it to be so? The answer lies in the psychological mechanism of denial. Unless directly and experientially confronted with the reality and horror of companion animal overpopulation, many people can blind themselves to this reality despite ample information input. They have a *need not to know*.

The same principle operated with respect to abusive animal experimentation. Until confronted with PETA's film, "Unnecessary Fuss," which documented clear abuse in a major and prestigious biomedical laboratory, many people just couldn't acknowledge the reality that was described to them again and again. The film was so effective in graphically demonstrating the horrible conditions that it led to a national hue and cry, and the laboratory was soon closed down.

Likewise, the companion animal overpopulation problem remains an abstraction in the minds of many people who, for a variety of reasons, deny the nitty-gritty details and continue to protest euthanasia. Euphemisms such as "put down" and "put to sleep" also contribute to opposition to euthanasia. Since everyone knows that the animals are being killed, and since we can't actually say it, it must be wrong. We should call it the way it is, and then justify its necessity.

Finally, in regard to euthanasia, the parallel to euthanasia in humans—a highly controversial subject—should be resisted. Although euthanasia in humans may be desirable in certain instances, parallelism with animals is not to be found and only obfuscates the issue. Because of cognitive differences between human and non-human animals in this area, anthropomorphism, sometimes desirable, should be avoided. Obviously, anthropocentrism, as usual, has no place, since what must be considered in decisions pertaining to euthanasia should reflect what is good for the animals, not the person.

What are some of the psychological influences derived from identification that alter attitudes toward neutering? I will not cover those influences which facilitate the decision to neuter, but rather only those which impede or obstruct:

Human males who are unsure of their manhood can be expected to show resistance to the idea of castrating their male animals. This would be especially true if the animal represents a narcissistic extension of the person, allowing the man to feel gratification at the aggressive and sexual behavior of the animal. This point was made earlier today. This state of affairs is carried to an extreme in the current pit bull crisis, where innocent animals are being blamed for severe disturbances in the humans who have used and abused them in the service of a person's need for self-esteem, elevation, macho self image and outlet for uncontrolled aggression. This dynamic seems to influence the decision to neuter male animals much more than a similar concern in women might influence the decision to spay.

In 1985, a *Dog Fancy* readership survey showed that human cohabitors of both female and male dogs in the same household were four times as likely to neuter only the females. This could reflect the myth that female dogs, since they actually have the babies, are the culprits, and that male dogs can be left sexually intact. This, of course, is nonsense, as unneutered male dogs can produce countless numbers of puppies. This same reaction pattern is found in men who take no responsibility for impregnating the woman they are sexually involved with, and blame her as if he had absolutely nothing to do with it. People who behave this way are either ignorant, very concrete in their thinking, or are rationalizing underlying concerns and conflicts.

The Jewish tradition of *Chalaka*, forbidding orchiectomy, may also play a role.

Freud's concept of castration anxiety, although widely refuted and often applied in much more subtle and seemingly unrelated areas, probably applies here if there is sufficient identification with one's companion animal. In more general terms, castration anxiety, often experienced as the fear of being maimed, might account for people's resistance to any major operation, whether it be neutering or anything else. The entire notion of pain and some people's morbid fear of it might cause them to resist any elective major surgery for their pet, despite the fact that, according to experts such as we have just heard, the amount of pain perceived by the animal is minimal. But fear that something might go wrong with the operation, the possibility that the animal might die, can strongly influence some guilt-susceptible people from taking responsibility and cause them to evade the issue.

The guilt that *should* be experienced from the awareness of the misery in store for the generations of progeny can be guarded against in some people by ignorance, both natural ignorance and forced ignorance, denial, or the narcissistic feeling—all too prevalent in this day and age—that I care only about *my* child, *my* pet, etc. It is renunciation of responsible stewardship over companion animals.

Identification will also explain the belief, this time perhaps more common in women than in men, that every female companion animal should have at least one litter. Physiological considerations aside, the only validity to this belief is anthropomorphic identity confusion between the two female members of the human-animal dyad, caused perhaps by feelings of incomplete femininity on the part of the woman, or by unconsciously using one's female companion animal to provide one with either a family one never had, or a larger family, or siblings for an only child. Since animals, unlike human females, are conceptually unaware of their femininity, this belief is merely self-serving, is not need-fulfilling to one's companion animal and neglects animals in general.

Likewise, the desire to show one's children the wonder of birth by allowing them to observe the family pet give birth should not be necessary in a household where humane education is emphasized, and may serve to compensate for a general absence of reverence for life in that household.

There are psychological mechanisms at work that do not depend on identification:

*Rationalizing* accounts for the belief that somehow humane societies and shelters will take care of the progeny. In reality, the progeny will probably be killed.

*Grandiosity* may account for the conviction that "I am different; I will manage to place the babies in good homes." Most people cannot.

I*nability to think long-range,* often a sign of emotional immaturity, could explain the avoidance of consideration of the question of what happens to the progeny when they grow up.

Many people are *resistant to change*, finding it frightening and threatening. They would be refractory to changing previously-held beliefs about neutering, despite new input and greater information.

*Risk–avoiders* would endlessly put off finally bringing in the animal. Intolerance of negative feelings, such as doubt, anxiety and worry, would prevent some people, often dependent types or people prone to depression, from neutering their companion animal.

*The need for certainty* on the part of some people would serve as an obstacle, both in terms of the small possibility of an unsuccessful operative result, as well as the need to feel that the indications are 100 percent without controversy. People who cannot make up their minds, who ruminate a great deal, who need things in perfect order, may be suffering from an obsessional disorder and could find great difficulty in making the decision to neuter.

Neutering does involve a certain amount of unpleasantness in terms of expense, risk, inconvenience and time commitment, and people who are pleasure-oriented and tend to avoid unpleasantness, such as that ever-proliferating group whose basic ethic is "feeling good about myself" would not be inclined to regard neutering their companion animal as fitting into the general scheme of self involvement. But fear of one's own anger or a considerable degree of underlying anger might explain the person's unwillingness to gently confront the subject with a friend and encourage the friend to neuter his or her animal.

Finally, the knowledge of pound seizure and the realization that, in many instances, the shelter is not a safe haven for animals, but rather a way station on a route to suffering in a laboratory, might cause people to avoid thinking of euthanasia, neutering or anything else that involved use of a shelter or

## Attitudes Toward Neutering and Euthanasia

humane society because of basic and often unconscious mistrust, and this, indeed, is a tragedy.

**MULLEN:** Dr. Cohen, thank you so very much for your perceptive and enlightening view of psychological and religious blocks to neutering and euthanasia.

**MULLEN:** We will now hear from Ingrid Newkirk, on the viewpoint of animal rights advocates concerning neutering and euthanasia.

**INGRID NEWKIRK:** It's nice to know that we are normal, isn't it? You and I are supposed to be the animal nuts, and Murry's talk puts things in a much better perspective.

Just as an aside, while I've been here, I saw two things on the TV which were kind of interesting, and one is: in New York you have a commercial, I think it's put out by the Greater Council of Churches or some august religious body. When Murry was talking, I remembered it. It's a poem, and it says something like: "Are you greater than the sun who shines on everyone?" Then it goes through a litany of races, and ends with, "The sun does not discriminate." And the interesting thing to me in watching it, all the background is animals. There is the sun shining on sparrows and the sun shining on all these different animals, and I thought we could just use that commercial right now.

The other thing, while Murry was talking about Israel and the strychnine poisoning, I remembered Nina Natelson, of CHAI, who joins us in thinking that it's very hard to take a vacation when you're in animal rights or animal welfare. Wherever you go, there it is, all the animals say, "Oh, here's another one. Let's remind them of our plight." On TV here, I have now seen that they're plugging holidays in Israel very hard. It would be a good idea if anybody feels like it, to write to the address—it's, I think, the Israeli Tourist Board—and remind them that it is very disconcerting to be in Israel, as Nina says, and see all the strays on the beaches, the starving kittens in the streets, and to know that they're being strychnine-poisoned. That might be something of use.

I'm supposed to be talking about the differences between animal welfare and animal rights, and with the perceptions of what happens in euthanasia, spaying and neutering, and so on. I guess the basic thing is that many of us came into the animal rights movement from backgrounds in animal welfare, as I did, and our philosophical evolution was very gradual. We started off caring about animals with whom we were familiar, and we saw the extent of problems with companion animals, with working animals, and then we began to have opportunities to branch out and extend our caring attitudes into other areas—other forms of abuse and exploitation. And our knowledge of those things grew as we ran into other people who told us about other books and films and situations they were involved in. So for us, it's easy to understand euthanasia and sterilization.

But many people in animal rights have never worked in animal shelters, as you know, and they came to animal rights by a very different route. They came in very directly and boldly, and they sort of just dropped into the philosophy, the belief that animals, the other animals, don't belong to us, that they don't exist simply for us to use. They weren't sitting there for millions of years waiting for *homo sapiens* to come along and go shopping for clothing and food and what have you. So those people are very lucky in a way, because it dawned on them, something that for others of us it took a long time to come to grips with.

Harriet Schleifer, who is an ecofeminist in Canada, was telling me how she went fishing with her father when she was five years old. It was the first time she had seen fishing, and she instinctively, instantly knew it was wrong that that fish was dying, and that this was not an act that she should be part of, or that her Daddy should be part of, and she told him that this was wrong. And I envy her. I've heard that from other animal rights people, that when they were little children, they saw the chicken's leg on the plate, and they knew what it was, and they didn't like it, and they said, "No."

But these animal rights people often lack animal welfare experience, and therefore, those of us who know about animal welfare shouldn't judge them too harshly, if sometimes they form opinions on domestic animal issues and on the conduct of humane society-type people from a purely philosophical-intellectual standpoint, without the benefit of inside experience. They don't have it. It allows them the luxury, really, unknowingly to brush aside these really tough practical problems that animal welfarists deal with on a day-to-day basis. In a way, it sets animal rights people in a kind of aloof position from animal welfare issues.

All of us share the common goal that we long for a time when animals are not ours, when animals are able to live their lives free and whole without interference, but that is not the case now, and sometimes we have to remind people in the animal rights movement that that's so, and that very immediate solutions to very horrible problems in a very unnatural world are necessary.

There are four main criticisms that I hear in animal rights about the perception—and I think it is just the perception—of animal welfare issues. They are:

**1.** Stray animals should be left on the streets; they don't belong to us to capture, to take back to the shelters and to inevitably dispose of.

**2.** It's unnatural to impose our spaying and neutering criteria on domesticated animals.

**3.** Euthanasia, of course, is morally indefensible.

**4.** The keeping of pets is wrong.

I guess we've gone into a lot of these things. I won't belabor it.

As to the strays issue, all we have to ask is: please ride along on the truck; please go to the shelter; please visit the pound; please spend time before you offer your opinion, and see what happens to animals in the cold, see what happens to them in the heat, see what happens to them in the alleyways and in the hands of juveniles and others; and then please tell me if your opinion is the same. I think there is a perception—again, it's just lack of experience—that that fat golden retriever wandering down the rich neighborhood in Chestnut Street is the one the warden was after. And if that *is* the case, the person who owns that fat golden retriever should have the opportunity to come and get him back from the shelter and pay a fine and keep him home, because he may end up in a laboratory if the dog warden doesn't get him first, and it's an education for the owner.

He may be hit by a car, all sorts of things may happen to him that the average person doesn't have the opportunity to see. But by and large, of course, everyone who works in shelters knows the truck is not going out looking for that fat golden retriever in a rich neighborhood. It's the ones they're scraping up off the street, going under the abandoned houses for, and the ones to whom all sorts of other things happen.

Secondly, pets. Now to me this is a very important point, it's a touchy issue. Animal rights people, of course, are opposed to the keeping of pets, but the pet issue is important because the fact that humans made animals "pets" is at the root of all of these problems. We wouldn't have to deal with the sterilization issue, with the euthanasia issue, with the animal shelters, with the surplus population, nothing, if we hadn't decided that we wanted to keep animals. Animal rights people find the use of the word "pet" repulsive. To us, it's like the word "doll" or "baby" that's used to describe women. It connotes that the individual is an object, a lesser individual, a frivolous creature who exists to amuse or to please an important person like the utterer of the word. We think of domesticated dogs and cats not as pets, and language is important, because as a culture evolves, its language reflects the amount of respect it has for various individuals within it. So we think of the animals, domesticated animals, as companions, as refugees. Not that we've chosen them voluntarily. We think of them as individuals with independent worlds, displaced by the very human society that brought them into being. And, as refugees, they must be sheltered, they must be fed, they must be rescued from cars and rock-throwers and theft, and all the things that we have created for them.

Animal rights people aren't really a problem to animal welfare people in other areas. They don't buy or sell animals or breed them, because they think of those activities as associated with slavery. Animal rights people try to treat the animals not as adjuncts to our lives, but as a part of our lives. We correct ourselves, or try to, if we find ourselves thinking of the dog or the cat in our home as a child or a surrogate human being, when we know that, unless they're puppies and kittens, they are adult, thinking individuals who should be respected and protected, and whom we have invited there to share our lives.

We tend to give the animals we live with real names too, rather than the names of cocktails or royalty or objects, although they're sometimes stuck with the vestiges of our past. We try to shake the use of the word "it" to describe a member of another species who clearly has a gender, just as we do. They are not objects. And if we catch ourselves talking to a dog or a horse as if he or she were a child or a bit retarded, we try to correct ourselves, and we try to look at them a little less condescendingly.

We pay attention to the manner in which we respond to the animals in our homes, which requires a lot of discipline, because it's not the normal way that people relate to animals in their homes, and it's usually not anything we've been taught. It requires a rethink. We find ourselves abandoning old habits. We don't eat while they stand there salivating. We don't feed them last, after the family has eaten, any more than we would do such a thing to a human guest or a friend in our home. We don't teach dogs tricks. Why should they have to beg for things? And, in general, we just try to examine, re-examine our relationship with animals and be more understanding. We don't always succeed. There are always gray areas, but we always have to try.

Pet ownership goes against the animal rights philosophy, and we have to work to phase it out. Now, this frightens a lot of us, because we all adore dogs and cats; we're very fond of them; we don't think we could live without that companionship sometimes. It's rather pompous to think that they couldn't live without ours. If you watch something like Jane Goodall and the wild dogs of Africa, where those dogs have never been domesticated, you see that their lives are complete, they're full. They have families, they have things to do, places to go. They know about more things than we do. They know about the seasons and the earth and their relationship. They know how the weather is going to turn out without reading the Farmer's Almanac or switching on the five o'clock news.

Some people say, "What will we do with all these animals if you don't have pets? If the animal rights people had their way, we wouldn't have pets." Well, that's right, but like everything else, things get phased out. Nothing happens overnight, but we have to move towards that, we believe. It's the same thing with: "What will you do with all the animals who are bred for meat?" I don't think that's been a justification for continuing to eat animals. If everybody stopped eating animals tomorrow, we would have this huge problem on our hands. We would have what they slaughter in one year sitting in front of you. But that's not a realistic problem.

If pet ownership declines and eventually disappears, it will be fine. It doesn't mean our lives will be barren either. Symbiotic relationships with other animals are currently possible. Feral cats go into the gorilla cages of the zoo. I think they perform a wonderful service for the gorillas. Feral cats and people. Squirrels and birds. Even dolphins enrich the lives of human beings now. Dian Fossey and her gorillas. No one had to take Digit home, put him in the kitchen. More would certainly be possible if we stopped being a species that inspired fear in our fellow animals. Cranes sit on the backs of buffalo; the deer and the squirrels eat together in the grass. All sorts of animals have wonderful relationships together. The problem is that it's the sight of human beings, it's the sound of the human footfall in the forest, that makes all the animals scatter and run and hide and collect their young and go underground.

Up to now, pretty much the only relationship we've been able to get towards is a relationship of capturing them and owning them and bringing up their young in captivity before they'll trust us. So we have to move towards a better day. This re-examination of our relationship with what we call "pets" today is very important. But just as animal welfare people must accept the broadening of our concerns to include *all* animals, so must animal rights people understand that animal welfare people are a very vital link in the bridge from not caring at all about animals to beginning an ethical evolution of concern for animals. And while animal rights people may dwell on what *should* be, on the utopia of our existence at some point in the future—if we have a future—animal welfare people have to be a part of the practical advancement of that ethic. Reducing the numbers of the unwanted by promoting sterilization is necessary, as is giving the gift of a traumaless death to those whose options for a quality life have been reduced to nil.

Both animal rights people and animal welfare people have very serious obligations. Animal rights people should be practical; animal welfare people should be forward-thinking. Humane societies need to open their hearts and minds to vegetarianism. After all, it makes absolutely no sense to spend one's day rescuing some animals and then support the hideous confinement of other animals when you go home to dinner. And animal rights people on their part have to recognize the debt we owe to what you described as all the people in the trenches.

**MULLEN:** Thank you very much, Ingrid.

## Dr. Tom Regan

**MULLEN:** We're now going to hear from one of the leading philosophers and certainly one of the most prolific authors in the animal rights field, Dr. Tom Regan—and also, in that same person, from a champion of animal rights—who did not enter the field via the animal welfare route. It will be interesting to see the contrasts and similarities in the points of view of Dr. Tom Regan and our previous speaker.

**DR. TOM REGAN:** Well, Ingrid's always a hard act to follow, as you know. In fact, ever since I've been here, which was only this afternoon, people have been talking about her momentous presentation of yesterday, and I regret very much that I didn't have an opportunity to hear it.

I do want to thank everybody who has given me the opportunity to be here, especially you, Marjorie. I think you must deserve the major credit for getting me here to talk about how it was that a person who did not start off the way that Ingrid did, got to where I am.

I was far removed from any of this stuff when I was a young person. In fact, at one point in my life, I worked in a butcher shop. I was a butcher's apprentice, and there just was never anything going on in my mind at all about ethics in relation to animals. Over a period of years, of course, I changed my mind. In fact, one of the things that I find rather interesting is that sometimes the heart leads the head, and sometimes the head leads the heart. That is, I think in Ingrid's case, she came to it through feeling, and then the philosophy came along later. I worked a different way, because the philosophy was there before the feelings were there, a lot of times. I find that the more I understand things, the madder I get, in fact, and I actually do think that the more I understand things—I'd like to believe this, anyhow—the more compassionate and the more loving I can become. So I think that it's not always the case of feeling first and thinking later—sometimes, the reverse can happen.

I was reminded, when Murry was talking about some views about animals, and how they don't have souls and all the rest of that, I was reminded of a lovely line by James Dickey in a poem called: "The Heaven of Animals." I commend James Dickey as a poet to you, and "The Heaven of Animals," in particular. It's just this one little line where he says: "Having no souls, they come anyway."

## Attitudes Toward Neutering and Euthanasia

I've been encouraged by the things that I've heard since I've been here. Certainly I think Ron Scott's work deserves our special commendation, because if there is a way of taking this message in an organized way into the school system, what Ron's doing is very important. I was also struck by Sheri Trainer's remarks about involving other constituencies. When I talk to activists, I say things like: "There are a few axioms of activism. The first is, 'People are busy.' The second is, 'Don't call a meeting, because the people who are busy aren't going to come.' And so the third thing we learn is, 'Go to theirs.'" What we have to ask is: "Who are those people who are meeting?"—and then take advantage of the fact that they're going to be in a room together. We need to create a program and go and show it to them. This works! Time and time and time again. I know in our area when we call a meeting, the people who show up are the people on the program. You've had that experience, I'm sure. But if we go to a meeting of the Luther League or the Wesley Fellowship or the adult Sunday school class or something else in the religious community, and if we go to the Rotary Club or the Kiwanis Club or the Lions Club or the Garden Club, there's an audience. A *new* one. People are meeting all the time, and all we have to do is try to figure out how can we take something of what we want to teach them. Remember now, we've been doing this, you and I, for years, and we can't get all that information in thought and feeling and experience of years into them in thirty minutes. But we *can* get *something*, and we have to kind of target the audience. Who are these people? What can we leave them with?

So all these things, I think, are terribly encouraging to me, and they've been the "up" side of being here, trying to think about what I regard as this terribly difficult problem about the morality of euthanizing companion animals.

One of the things that makes this very hard for me is that usually when I think about an animal's misery, there are identifiable people whose very existence makes my blood boil, and so I have no trouble thinking about it. There's Gennarelli, for example. I just have no trouble thinking about the question of the use of animals in science. Or there's Colonel Sanders, and I have no trouble thinking about it. But in the case of euthanasia, the front-line people here are caring and loving and good and decent people, and so it's just very hard to even think that there might be something wrong with its being done. Not in the case of Gennarelli, not in the case of factory farming, not in the case of rodeo, not in the case of all these things, but in this case, there is this question that we need to ask ourselves, whatever the answer.

Now, in the animal rights philosophy, that whole challenge of trying to think this through is really exacerbated, I think, because the animal rights philosophy takes the notion of the individual as of fundamental moral importance. The way we think about it is something like this:

First, let's think about it in the human moral community. When I think that I have rights and you have rights, part of what I think is that I don't exist for you. I'm not yours. And even though it might be useful to you or beneficial to you or therapeutic for you or whatever, to treat me merely as a means to your ends, because you've got superior power or wealth or knowledge, that doesn't give you the right to do that. In fact, it would violate my rights if you did that. You're not showing respect for me—simply to treat me like a pencil or an eraser or a car. I'm not a thing. Well, in the case of other animals, then, the animal rights philosophy certainly thinks that this same principle—respect—applies, and what we must struggle to do in our treatment, in our interaction with other animals, is to treat them with the same kind of respect with which we struggle and strive to treat one another.

Now, the question is, how can we be doing this when we're killing animals? When we look at Gennarelli, we know he's not doing it. When we look at factory farming, we know they're not doing it. And when we look at rodeos, we know they're not doing it.

Now, in the case of euthanasia in a shelter, are we doing it there? Are we showing respect there? I think there's one clear case where it certainly is true, and then there are the more challenging cases. The one clear case where I think it certainly is true is where you have the terminally ill companion animal whose quality of life is so impaired and so diminished that death is a merciful release from this condition. And sometimes, of course, we're in a situation where the animal is teetering on the edge, not quite there but is soon going to be there. It's a gray area, but soon that animal is going to be there, and so we question again: "Don't we owe it to the animal, as a sign of respect for the animal, to prevent that from happening to the animal?" I think those cases are the paradigms, the clearest cases where euthanasia shows respect.

But a case where the animal either is terminally ill or teetering on the edge of getting in that kind of condition, is different from where we have a perfectly healthy dog or cat for a companion animal. It just so happens that in the lottery of life, this companion animal isn't wanted by anybody and hasn't been claimed by anybody. Now, how is it that, if we kill this animal, we are acting in a way that is consistent with the animal rights philosophy?

Notice this—and this is very important: A lot of people, when they try to defend the intentional killing of healthy animals because they're unclaimed and unwanted, argue that this is necessary in terms of social benefits. And they'll talk about any number of things. They'll talk about: "Well, if we were to release these dogs and they were to multiply ..." and so on, and they'll talk about the threat that they pose to human health, and things like that. Well, what's important to realize, I think, is that that changes the focus; that gets us off the question we're really interested in. The question is not: "Is society going to benefit more or is society going to benefit less from our killing healthy, unclaimed and unwanted animals?" That's not the animal rights question. The question is: "How is it that we're showing respect for this animal?"—not: "What's going to happen to society?" How are we showing respect for this animal if we kill, intentionally, deliberately, and lovingly, this healthy dog or cat? Now, the best argument I think that is available to an animal rights person in trying to defend this kind of practice, the best one I can think of—and it may not be right, but it's the best one I can think of now—is the following: *I think that if animals do have a right to be treated with respect*, then obviously, in my view—and I think you'll share this with me—*they all have that right equally*. It's not like some have it more than others, but they all have the same right, and they have it equally. And I think what this implies is that all should have an equal opportunity to have a good life. When it's not possible, for them all to have a good life, what we have to do is to try to give each one *the same opportunity* for a good life. We have to try to figure out some fair policy for doing this.

Now, in an imperfect world, it's hard, I think, to create equal opportunities that don't have some degree of arbitrariness. And we are in an imperfect world. But a possible way of trying to do this, I think, would be to have a fixed amount of time, a specified amount of time available to any unclaimed and unwanted animal, to have the opportunity not to face death but to find a new home. Now, if we extend that period

## Dr. Tom Regan

of time beyond a time, again, which might be arbitrary: three days, four days, five days—if we extend that period of time indefinitely, then I think the great challenge and the great problem is, that we're not treating the other animals who can't come into this safe haven, however temporary it may be, with the same kind of respect that we're treating the animals there. That is, "first come, first served" doesn't mean you have a right over and above those who are waiting to get in.

If we're to treat those animals outside the door fairly and equally, and give them an opportunity to find a good home, the great tragedy is, we must make room for them, and that means killing those who have not been claimed and those who are unwanted. And again, what I mean by this, mind you, is: I don't think this is a wonderful event, that this is something we should take tremendous pride in: "Aren't we a wonderful species?" I mean, it's a great tragedy. But again, if we're to treat animals equally, with equal respect, we must think of the animals outside the shelter, not just the animals inside the shelter, and if we're to give that animal outside that shelter an equal chance, a chance equal to the animal inside the shelter, then we've got to get that animal outside the shelter inside the shelter.

The great tragedy is, there seems, at this point in time, only one way to do that. I believe that this is a position that would justify active, intentional killing. I have trouble with the word, "euthanasia" here, because the paradigm for euthanasia is when the dog is dying. This is a healthy animal, and I have trouble calling this euthanasia. But I do think it is a good death for this animal. Under the circumstances, it's the best we can do. It's also, I think, not only the *best* we can do. Given the principles of an animal rights philosophy, it's what we *ought* to do.


**MULLEN:** Thank you very much for agreeing to address these very difficult issues.

I know that some of you will have questions for our speakers.

While Ann Free is coming to the microphone, let me take this opportunity to thank the volunteers who are photographing, audiotaping and videotaping our conference proceedings: Hilda Daily, Edward Ashton and Ron Scott.

## Attitudes Toward Neutering and Euthanasia

**ANN FREE:** My question is to any and all, but particularly Ingrid. This labeling, "animal rights, animal welfare," is causing a polarization, I think we'll all agree to that. Is there a solution? Should we not do something about this semantic trap in which we are caught by coming up with something entirely different? And the second part of the question is: You are certainly in that middle thing—rights and welfare—so are you going to go through life a "rights/welfare"? So, in other words, is there something else that would be everything? "Protectionist?" That makes you look like you're the Great Father Above. So, is there any way out of this semantic jungle? That's my question, and I think then we might have more peace and unity.[28]

**NEWKIRK:** I tend to think, rather, that you can have a positive polarization, or a connected polarization. I like the debate about animal welfare and animal rights because I think we have to concentrate on our commonalities, and not on our dissimilarities or how we've failed to come together, but really our common link. In your writings, you've used "humanitarian" and "protectionist", and I think those are wonderful words. But I do think we have to discuss these things, because we've all got to move together; because we're all moving in the same direction.

**DR. MURRY COHEN:** I'd just like to add a few thoughts to that. I've grappled with this question. I think that it's an important question because "animal rights," although much more accepted and embraced than say, two years ago, when you would be automatically branded a lunatic, if you admitted to such a thing, is now becoming (less so in New York City but more so across the country, in my experience) sort of a mainstream issue. But it still strikes a lot of people as a bit weird: "Why be concerned about animal rights when there is child abuse?" and all the things that people ask us why we're not involved in, so that's kind of tricky.

"Animal welfare" is also tricky because there are radical animal rights people who will treat you with scorn and disdain if you express interest or views in animal welfare, for the reasons already mentioned. They haven't been out in the trenches and, actually, *I* haven't been out in the trenches—unless you want to call medical school a trench, which actually, it is.

"Animal liberation" is kind of tricky because that sounds like you're a follower of a particular person.

I have found personally that "animal advocacy" is a term that everybody can agree with. I mean, it's sort of neutral, and everybody feels comfortable with it, and it communicates what you want to communicate, so whenever I'm at odds with knowing who I'm talking to and dealing with, I'll use the term, "animal advocacy," and then I might get more specific. But that's one way out of it.

**GRETCHEN WYLER:** Don't you think that separatism is important now because, when the animal rights people say they're animal rights, right away, they mean, don't eat and wear animals. That doesn't mean the animal welfare people as an entity are not very valuable in our society, but clearly, if someone says "I'm in animal welfare," it means they are not into animal rights. I think it means that the separation is important.

**ANN FREE:** But maybe they are.

**WYLER:** Well, then they'll call themselves, "animal rights."

**FREE:** I know, but [Ingrid]'s both.

**WYLER:** So am I.

**MULLEN:** Thank you so much for an extraordinary presentation.

# Legislation: Past and Future

**MULLEN:** Our next speaker is Elinor Molbegott, an attorney who is General Counsel for the ASPCA and its Vice President for legislation. Elinor is also a former director of the New York State Humane Association, and she still continues to serve us very well and very actively, though she hasn't been on our board for a couple of years. Each time we call her with a question directed at her field of expertise, she is unstintingly generous with her time and help. Elinor Molbegott will speak to us about legislation affecting the issues that concern us here today.

**ELINOR MOLBEGOTT, ESQ.:** It is a pleasure to be here today. It's interesting, I was listening to Ingrid before, and I said to myself; "You know, I never heard that story about the chicken leg." And that's exactly what happened to me one day. The chicken leg was on the table, and I said: "What *is* this?"—after years and years of eating it. And this was a long time ago, and it was the last time it was on my table.

I'm speaking today about legislation. But before I go on to talking about some of the issues that we deal with relating to the focus of this conference, which is the overpopulation of dogs and cats, I would like to say that I was interested to see so much emphasis on humane education at the conference, because, in reality, we're never going to get the good legislation through, and enforced, until the public is educated about the need for such legislation; is educated about the issues concerning the overpopulation of animals and about all the indirect issues that have somehow come together with the overpopulation issue, which I will get into.

Legislators are just members of the public who happen to represent us, and when we go to Albany or Washington or to any local area, we find that the same questions that the public asks, the same ignorance we see from members of the public

Elinor Molbegott, Esq.

on the issues, we see from the legislators. We usually have to start at ground zero.

One of the key issues, for example, that affects overpopulation that's pending in the New York State legislature, that's pending in other state legislatures and that has passed in some, deals with spaying and neutering of animals by shelters. In New York, we have a bill introduced by Assemblyman Hinchey, that would require animals to be spayed or neutered before they leave the shelter. And if this isn't done, then the alternative is, the shelter could require the adopter to sign an affidavit that they're going to have the animal spayed or neutered and leave a deposit to provide greater assurance that this procedure will be done.[1]

When lobbying for this legislation, the typical question asked is, "What's wrong with breeding animals, dogs and cats? Why would you want to have such an operation done?" Office after office, this question was asked, which makes me realize that, before any legislation is passed and taken seriously, the public must be educated, so that these very initial questions are not asked, because once you've gotten the legislator on your side, you then have to convince the many other legislators who may not agree with your point of view, once educated to it, or feel for one reason or another, even if they do agree with your point of view, that their political philosophy is such that they want to take a hands-off attitude. So there are many, many points that come into play when legislation is being considered, that decide the substantive provision of a given bill. And if you have a completely uneducated audience before you, you get nothing accomplished, and that's why legislation probably more than anything else is so frustrating: because legislation comes after education, and we all know how long it takes for people to become educated on the issues that concern us here.

When I was first asked to speak today, I thought to myself, "Well, overpopulation: What bills are pending, or what laws exist that deal with the overpopulation issue?" And very little comes to mind. I thought of the bill introduced, as I just mentioned, by Assemblyman Hinchey, that deals directly with spaying and neutering dogs and cats at shelters. And I thought of one other, and that is the dog license differentials that exist in some areas. In some states and localities, when you get a license for your dog—which is required all over—if your dog is spayed or neutered, you pay a lower license fee

## Legislation: Past and Future

than you would pay otherwise. Such bills have to be encouraged, because that's a way of encouraging people to get their animal spayed or neutered. But then I said: "Well, I'm given quite awhile to talk about the subject, and if that was it, then I would be sitting down and everyone would be leaving, and what else there is to say, because that's really what exists in this area." But then I gave it a little more thought, and so many other issues came to mind, so many issues that we work on—that we work on for a variety of reasons, the overpopulation problem not being foremost—but they all do relate to the overpopulation problem.

A key issue is pound seizure, which I know you've heard a lot about at the conference. If one assumes that one of the arguments used against pound seizure is that people will be reluctant to leave animals at a shelter, if they know that the shelter might give an animal to a laboratory—if one accepts that argument—then one has to realize that pound seizure directly impacts on the overpopulation problem. Because we have an exacerbation of the overpopulation problem if people are going to leave strays on the street, or perhaps, in cases where they can't keep their own pets, they decide to leave those pets on the street where they might, in their minds, be better off than in a shelter and then in a laboratory. I believe that argument. I do believe that people leave animals on the street and abandon animals on the street, when they perceive the shelters in their areas to have inhumane practices, pound seizure being one of them. But I have good news, at least on this issue. In New York State this year, a law was enacted which really tightens up the loopholes in our pound seizure law.[8]

Years ago, the Metcalf-Hatch Act was repealed. This act, when it existed, stated that if the health commissioner requisitioned animals from publicly supported pounds, then those animals had to be given over for research, and many shelters complied with that. And, as a result, I do believe many people left animals on the street or abandoned them. When that law was repealed, there was a void left. Some of that void was filled with legislation to restrict what could be done with shelter animals outside of New York City, in certain municipal shelters. The way the law was written, though, it was subject to a variety of interpretations, unfortunately, and it was never enforced. So, while it *did* imply that dogs and cats at municipal shelters throughout the state may only be adopted for use as a pet or euthanized, the department responsibility for enforcing that law, the state Department of Agriculture and

Elinor Molbegott, Esq.

Markets, did not interpret the law that way. So there was a void. There was also a void because that law did not apply to private humane societies and it didn't apply to the City of New York.

Now, two laws were passed this past session in Albany. One of these laws applies just to New York City, and the other law applies to the entire state. In essence, though, they do say that no shelter animal in the state may be transferred for research, testing, experimentation, teaching or demonstration, whether it be a public humane society or a private humane society. So the option is no longer there for the private humane society which, to get an extra dollar, has sold animals to laboratories voluntarily, even though they haven't had the mandate to do so, for several years now.

So we've come a long way, and I think that, when the public is educated to know that this is the way the law currently is in New York State, there will be greater trust in our shelters.[19] Of course, we have a way to go on this issue of pound seizure, even in this local area. We should help our sister states, at the very least. The laboratories in New York State may just go to another state and get the animals there, since they're not going to be able to get them from the shelters in this state. So we must work during the next session on legislation that will prohibit research institutions in New York from accepting any animals that originated in a pound or shelter anywhere, and that will be one of our priorities in the coming session.

Another issue that, again, didn't occur to me at first, but directly impacts on overpopulation, is euthanasia, which you've heard a lot about over the last two days, as well. Good news again: This session was very fruitful in terms of the legislation enacted, because after years of working to get a bill passed to make the destruction of animals in shelters more humane, a bill *was* finally enacted which bans T-61, which prohibits gunshot, unless it's an emergency, and prohibits the use of engine fumes, and facilitates our shelters' getting cheap sodium pentobarbital, which by many is considered the most humane means of killing animals at shelters.[8]

Hopefully, with public awareness of these laws, the stray on the street will be brought to the shelter rather than left on the street. The owner would bring his or her animal to a shelter rather than put the animal out on the street, and I think that, again, with public education, it will work, because when the decompression chamber was banned many years ago, people who never would bring animals before to the ASPCA have told me that they then did pick up the stray and bring the animal there, because they felt the means of killing the animal was more humane. So I think now, throughout the state, where more humane laws are put into effect, that people will begin to, perhaps more than before, bring the animals to the shelter and lessen the chances that those animals will just reproduce on the streets and/or suffer very inhumane deaths.

Another issue that I've been working on for years but didn't really think of as an overpopulation issue, is pets-in-housing. In New York City, we're fortunate enough, after years, to have a law that says where a tenant in a multiple dwelling has had a pet for three months or more with the knowledge of the landlord or an agent of the landlord, despite a no-pet clause in the lease, the landlord may not enforce that lease provision. And that has worked wonders. Instead of people lining up at the ASPCA, which they have done in the past, turning in animals simply because they've been threatened with eviction, we can now tell these people who want their animals: "You can keep your animal, because if your landlord does by some chance take you to court, he's going to lose, because this law protects you." So we've come a long way in New York City, although not as far as we need to go.

We need to go much further in the state and federally on this too, because I believe many, many more animals would have homes if people felt they could keep an animal without subjecting themselves to eviction. Unfortunately, we have many animals still being turned in, even though they're wanted. They join the ranks of an overpopulation of unwanted animals, because of laws that exist, because of constructions of lease provisions that make it very legally impractical for someone to keep a pet. So we have to change the laws that make it difficult for people to keep their pets.

On the state level, all we have is the public housing law in New York that protects people living in public housing projects, if they happen to be handicapped. But if you happen to be well enough to take care of your pet, you can't have one.

Federally, the same situation exists. If you're a senior citizen or you are a handicapped person living in special housing for senior citizens or handicapped persons, then you can have pets, but if you are a senior citizen living in public housing

Elinor Molbegott, Esq.

(it's subsidized, but it's not for senior citizens; the building itself is for people of all ages), then you're not protected by this federal law, because the law deals with the type of housing, as compared to the type of an individual in terms of age or handicap. So the federal law, if you're a senior citizen and you're handicapped and you live in housing for senior citizens and handicapped people—housing designated for such people—then you're protected.

Much more needs to be done to get us to a point where no person will be denied occupancy or be subjected to eviction solely on the ground that they have animals. And only when a nuisance is created should people be then subjected to a legal proceeding.

Unfortunately, we have a long way to go: We can bring countless people to Albany (and that's how the legislation was passed in New York City), who will explain their situation—how they've had a companion animal for years, and the landlord said nothing, and then suddenly, many, many years later when the building is going co-op, or they've had an argument with their landlord, the no-pet clause is being used against them. Legislators have understood this, but on the issue of whether a landlord should be able to have a no-pet provision in a lease, many legislators feel it's the landlord's property—the building is his property—and he should have the right to regulate that property in the way he wants.

So we may at some time, we hope, get legislation that will be similar to New York City's: If you've had a pet for awhile, and the landlord's known about it, he can't enforce that no-pet clause, because if it was considered important enough to him, he would have done something about it during the first three months.

Getting legislation to ban no-pet clauses altogether, to ban enforcement of them, is going to be a harder effort, because we go beyond the issue of whether a particular legislator believes that people should be able to keep their animals. We go to the issue of whether landlords should be able to control their own property and what happens in it. And unfortunately, you have many, many legislators who are landlord-oriented.

So, when we work on legislation, as I mentioned before, it's not only education, and informing a legislator as to what our point of view is, we then have another step to go in trying to

## Legislation: Past and Future

change the general philosophies, which is sometimes almost impossible to do.

But we've gotten the three-month law passed in New York City. I have hopes that we can get something similar passed on the state level, or that other communities will make the same effort and get local laws enacted where they're needed, because it is more difficult to do in the state as a whole. That is because, in New York State, we have a situation where there are some urban areas, and there are rural areas and suburban areas. Those legislators in the urban areas understand the problem. But if they're in a rural area, and many of their constituents don't have the problem, because they live in private houses, they just don't understand and they're not going to change their philosophy about what a landlord can do. So it's very, very difficult to get this sort of legislation through.

Pets-in-housing legislation has a tremendous impact on overpopulation because, as I said before, if you can't *keep* a pet, you're not going to *get* a pet. And if you have a pet, and you have to give up this pet, or feel that you do, and are in a position where you can't afford to move, and you give up your animal, then that just adds to the ranks of the unwanted, and the animal may end up getting killed.

Another issue that overpopulation impacts, is abandonment of animals, or rather, abandonment impacts on overpopulation. If our laws dealing with abandonment were stronger, hopefully they might serve as some incentive to these people to take their animals, at least, to shelters. I don't expect that such laws would provide incentive enough for a person to be a loving, caring companion to his or her pet, but perhaps if someone felt that they would be punished enough if they just turned their animal loose on the street[5], they would at the very least bring an animal to a shelter and thereby lessen the chance that that animal would be injured, or reproduce on the streets.

I have a list of the bills here that are now pending in Albany, and will be reconsidered during the session beginning next January. One of those bills deals with abandonment. It carries a minimum fine of $500 for abandoning an animal. It didn't pass, but last year was the first time the bill was introduced, and rarely does a bill ever get passed the first year. It's fortunate if it gets *printed* the first year. But it did get introduced. It's a very short bill. It doesn't seem on its face to do much in the overall scheme of overpopulation, but we need a combination of everything. We need pets-in-housing laws, we need more humane euthanasia laws, we need more laws dealing with pound seizure so that animals are protected, and only when all these laws get enacted and the public is educated, will there be a significant change.

Another whole area impacting on overpopulation is that of dog licensing laws. I can't tell you how often animals are in ASPCA shelters and shelters throughout the state that you *know* have owners. Not all animals in animal shelters were just born on the streets or were abandoned. Many of these animals were picked up, perhaps, by a good Samaritan who thought an animal was lost, but the animal might have just been out for a walk. An irresponsible person lets his animal out, somebody sees the animal loose and picks the animal up—a common scenario. Someone might keep the animal, and then bring it to a shelter somewhere else after awhile. There are so many stories behind each animal that ends up in a shelter. But I have to believe that many of the stories *could* have happy endings if those animals were identified properly.[5] We must have more stringent fines for people who violate license laws. These people might complain at first, but they'd be very happy if they got their pet back that was somehow lost. Unfortunately, in New York City, we have an $8.50 license fee and a $10 maximum fine. People not realizing the importance of the licensing law often violate it. They feel: "Well, if the fine is only $10, it's worth the risk." The chances of their getting caught aren't that great. So we need stronger licensing laws and we need better enforcement of these laws. Of course, you have to get people educated as to why they need a license, because until that's done, they're not going to buy one anyway.

We need large spay/neuter differentials, so that people will be encouraged to spay or neuter their animal. If they're not educated to do so by everything that goes on, sometimes you have to hit them in the purse and that's the only way that they'll understand and perhaps go through with the operation. Many people don't have a philosophical objection to spay/neuter operations; they just don't *bother* with it or they don't *think* about it, and we have to *make* them think about it. And there is the apathetic person who doesn't care about the overpopulation problem or feels that their animal will never go out unleashed and therefore will never create a birth problem, which we know is untrue. All good intentions aside, the animal somehow gets out one time. Sometimes you have to hit people in the purse.

We have to think of alternate way of identifying animals, such as tattoos.[5] This is for the protection of the animal from laboratories, first of all, but it's also a way that animals can be returned to their owners. It's a way that the animals in the shelter just don't become another statistic of the overpopulation problem.

I honestly believe many of those animals at the shelter could be wanted and are wanted. They're turned in, as I repeat again, by owners who want them but can't keep them because of housing problems; they're lost and they get there and they can't be returned because there's no way to know who the pet owner is. And that's tragic. It's bad enough that many animals in shelters are unwanted, but to think that so many of the animals could find homes, lessen the number of animals killed, is tragic. It's tragic that that's not happening because the laws are not strong enough, and our public education programs haven't gotten far enough so that people are obeying the laws and people understand the laws.

Leash laws, finally, are another issue which does not get a lot of interest among most humane activists. There's too much else going on, but again, so many of the animals ending up in shelters are lost by people who, with all good intentions, may have their animal off a leash for some exercise. The animal has gotten lost or may have been picked up by somebody even if the animal wasn't lost. And these animals also get added to the statistics of animals killed. We have to have higher fines for violations of leash laws. I'm thinking of one case that I wasn't involved in but I read about where again, that person pleaded that it was an accident, he never let his animal out before, but his animal got out and impregnated another animal. The animal had the litter, and the owner of the female dog brought, what they called in the case, a "puppy paternity suit" against the person whose animal had gotten loose. That person claimed in court that it was an accident, there was no negligence. The court did not accept that and actually ordered the owner of the male dog to pay paternity, to pay—

**VOICE:** Petimony.

Elinor Molbegott, Esq.

**MOLBEGOTT:** "Alimony," I was going to say, child support or puppy support, for a bed for the puppies, etc.

There's so much that needs to be done, and there are so many bills pending. In Albany alone, there are over 100 animal-related bills. I've just given you a summary of what exists, what the more important ones are that maybe you all haven't thought of in terms of the overpopulation problem, and maybe you haven't thought of as being a major issue. I know so often a license and a leash law don't engender much humane society support. And often, I find myself alone in Albany on the pets-in-housing issue, because for some reason it's not as sexy as others. Perhaps my interest in it is from seeing in the shelters so many animals being killed for reasons that don't have to be.

One case comes to mind just on that issue, and I'll close with that. An 86 year-old woman came to me one day, and she had had a pet for several years, a Samoyed. Her husband had died many, many years before, and she suddenly got a letter from her landlord saying: "Get rid of your pet or move." She was 86 years old and really not in a position to move from her apartment. She went to court. She came to me after the judge said: "There's a no-pet provision in your lease. The fact that you have had a pet for so many years is unfortunate, but I reluctantly order you out of your apartment unless you remove your pet." And if they do this to an 86 year-old woman, you can imagine they do it to everybody. We appealed her case, and unfortunately, she died before the appeal was decided, because it went on and on and on. But she did testify for us to City Council, and I think it was largely responsible for getting the legislation through. But these are the people and those are the animals. That animal would have been another statistic in a shelter.

When I talk today about pets-in-housing and license laws and leash laws, I know that many people have never considered them too important in the scope of experimentation, factory farming, wildlife matters. But lives *are* at stake. The overpopulation problem is caused by people failing to obey these laws. It's caused because we don't have strong enough laws, and it's caused because these laws are not enforced, and largely because we don't have enough laws, because the public is not educated enough. And so we need the humane societies, and we need the people who are concerned about animals to realize that while there are many, many other issues that we should all get involved in, that all of these issues that don't seem too sexy or too related to animal death, they are. They're very much related, and I hope we'll all in the future work together so that we can save some more lives.

**[See Update, pp. 240-241.]**

**MULLEN:** We will take a few questions, but before doing so, I would like to thank you, Elinor, for doing us the special favor of acting as our last speaker. That is never an enviable job.

**GRETCHEN WYLER:** First of all, my compliments. My question, I think, is to the assemblage. How is it possible that we've all met here for two days, and you have hit upon the most important things that nobody even looked at before in these last two days? I applaud you for bringing it up. I think she's very right, and I'm sure, the way she's articulated it, you all agree. It's not sexy, is it? Pets-in-housing, leash laws, and all of that, and you've hit on something. Even yesterday, when they said, "Who is responsible?" and everybody was coming up with different ideas, nobody mentioned pets in housing.

When I headed the volunteer program of the ASPCA, the stats on "given up because of housing clauses" were 32 percent. I used to go through and write it down for a few weeks so I could get a sense of the stats, and I would suggest (it's hard to do if you don't live in a metropolitan area), that we should make *much* more of that. I wish you'd run some stats of the ASPCA and *make* some noise about it. *Let* people get upset because, as you say, it's private ownership. But we could get a public campaign going, saying: "Forty-three thousand dogs are homeless because their owners were not allowed to keep them." Make some accountability out of it. I applaud you for bringing this to our attention.

**MOLBEGOTT:** Thank you, Gretchen. I should just say our statistics in New York City will probably be lower than elsewhere just because of our new law. It's not even that new any more. But the problem exists around the state, and the statistics we don't know and we probably never will, are how many people *would* come to the ASPCA and adopt a pet if they felt that they could keep a pet without being threatened with eviction, and that we'll never know, and we won't know that in the rest of the state. Our big obstacle is that the chairman of the Senate Housing Committee in New York State whom this bill goes before, is from Niagara and he doesn't understand or sympathize with the problem as much as we do, so we have an uphill battle.

**WYLER:** We should all start talking about it more.

**MOLBEGOTT:** That's right, we really have to, because I'm usually alone on this issue.

**WYLER:** Well, let's not leave you alone.

**MURRY COHEN:** I'd like to add my congratulations. I think this is the first aspect that we've heard about which focuses upon a demand function for pets rather than an oversupply function. Now, I do have one specific question though, because this influenced me when I was living in the Philadelphia area. Many of the houses that were being converted into condominiums and cooperatives in that area, that previously permitted pets under the ownership by landlords were now beginning to develop governing boards of tenants who were establishing new precedents outlawing pets. They were always grandfathered in. That is, any pet who was a resident at that time would be permitted to live out its life, but any future acquired pets would be banned by that governing board, and I was just wondering if that phenomenon was beginning to occur here in New York as more and more private buildings are going into co-op.

**MOLBEGOTT:** Very definitely. In fact, one of the reasons we were able to get our New York City pets-in-housing law passed was because so often, in order to get a vacant apartment, the landlord would threaten someone with eviction for having a pet because the landlord wanted to go co-op. Yes, the co-op and the condo conversions are very prevalent here, and the effect has been felt by the pet-owners. What we have, though, in New York City, is a law that does apply to co-ops and condos if they're multiple dwellings. The way our New

## Legislation: Past and Future

York City law reads, if once you have a pet and the no-pet clause is waived, it's waived, so that if you get future pets, it should be interpreted as allowing you to keep any future pet, as long as you've had your first pet for three months or more.

**COHEN:** In other words, this goes beyond the grandfather clause.

**MOLBEGOTT:** It goes beyond the grandfather clause. But we will only be home-free when we have a law that says you can't be denied occupancy or be subject to eviction on the sole ground that you have a pet. That's what we actually need.

**HENRY SPIRA:** It was good to hear Elinor discuss the impact of legislation on overpopulation. One of the things that I believe is that what we really need, in addition to what Elinor has suggested here is an integrated program. You know, while this is productive, I think one needs all the elements pulled together, including a tax on breeders, including a public awareness campaign—"Don't breed, don't buy, while animals die"—including letting the public know that if you take the animal to a shelter, the odds of it winding up with some Hollywood movie star, as a lot of these photographs show, isn't the reality. I think getting public officials involved, making spay and neuter not only cheap but readily acceptable and available, and going into the areas— I would tend to believe that if you have free spay and neutering, in the long run, it will be cost-effective to the taxpayers. But I think one needs an entire integrated program including hooking it up with the local vets, and really figuring out what all the elements are that you're going to need in order in the long run to be able to reduce the population. I think right now, most of the energies have really been on the killing of the animals, and I think the only way you've got to reduce any population is either reduce the birth rate or increase the death rate, and I think the energies really should go into reducing the birth rate. Thank you.

**MULLEN:** Thank you or your comments, Henry. And thank you, Elinor, for your comprehensive presentation.

# Closing of Conference

**MULLEN:** We're very fortunate to have in our presence a legislator who has introduced a bill[1] that would very effectively address the issue of reducing the overpopulation, and place added emphasis on prevention rather than euthanasia.

I'd like to introduce my assemblyman, from the 101st District, Maurice Hinchey, who was gracious enough to introduce not only a bill to mandate spay/neuter of all animals adopted out from animal shelters in the State of New York,[1] but also a beautiful resolution[1a] in support of this conference, that I hope all of you have had an opportunity to read.

I'm going to ask Mr. Hinchey to officially present New York State Humane Association with the resolution, and then I would like to present to him a citation which I'll read to you:

> *This citation for legislative service goes to Assemblyman Maurice B. Hinchey with our sincere appreciation for having introduced a resolution passed by the New York State Legislature in support of NYSHA's 1987 Conference on Pet Overpopulation and Stray Cats and Dogs, and in support of legislation that would implement the purpose of the conference: to help effect humane, responsible solutions to the problems created by the overabundance of cats and dogs found in virtually every community.*
>
> *This tribute goes to you with our gratitude also for your sponsorship of such legislation. Bill A.4968-A,[1] which you introduced during the legislative session of 1987, would standardize the policy of mandatory sterilization of dogs and cats adopted from animal shelters in New York State, and by means of requiring a substantial deposit of prepay-*

*ment towards the cost of neutering the animal would help ensure that such a policy was actually carried out.*

*The New York State Humane Association is deeply grateful to you for legislatively addressing the tragic problems caused by the uncontrolled proliferation of companion animals.*

**ASSEMBLYMAN MAURICE B. HINCHEY:** Thank you all very much. I know you're applauding for the issue, and I am very happy to have an opportunity to play a small part in advancing that issue.

Samantha said I was gracious enough to introduce the resolution. My being gracious really had nothing to do with it. Even if I had I not been inclined to introduce it, I'm afraid she would have seen to it that I did. Samantha is an irresistible force. We are neighbors; I represent her town, along with a number of others, the town of New Paltz. And I feel myself very fortunate to have an association with her. She is an educator of the first degree, a person of great energy who believes fully and completely in what she is doing. And because she believes so much in what she is doing, she is having an effect on everybody in the legislature, not just myself. She affects me most directly because I have to depend on her to vote for me or against me, so she has an extra leverage on me, but she affects everyone because of the level of her commitment. Her energy is incredible, and I want to enlist her in any cause that I am a part of, believe me.

The resolution recognizes the importance of this conference, and it's appropriate that we do that. We humans happen to be the dominant species on this planet, but that doesn't give us the right to treat other species in an unkind or inhumane or unreasonable way. We have responsibility, particular responsibility, because we are the dominant species, to all of the other species, and I'm particularly sensitive to the need to address the issue of endangered species and threatened species, because it is for those species of wildlife and other forms of life that we have particular obligation. But we also have an obligation to the animals that we call pets—dogs and cats particularly—and their proliferation in our society, and the real injustice that suffer: the hardships, the untimely deaths and the cruelties that they suffer as a result of overpopulation. And it is not only the right thing

## Closing of Conference

to do, but it's also in our interest as a species to control that population, keep it reasonable. And, as one gentleman said just a moment ago, it is even cost conscious. He said something about the fact that it makes economic sense for us to behave in that way, to subsidize spaying and neutering of animals so that we can hold down the population, because if we don't do that, then it will cost us a great deal more money in the long run.

It's with a great deal of pride that I come here to accept this citation from all of you. I'm very glad that my wife and I could make the trip down here this afternoon to be with you. I thank you very much.

**MULLEN:** We're very grateful that you could join us.

I think it's accurate to say that this has been an extraordinarily satisfying conference, and I would like to express appreciation to everyone who has been a part of it. Unfortunately, I can't thank all of the volunteers by name, but it should be known that we could not have organized the conference without the unpaid help of a large number of volunteers.

I'd like to commend NYSHA's board of directors, and to thank once more our exceptional speakers, people of extraordinary gifts and commitment. I consider myself very fortunate to be working in the company of people like you. I believe that, when we're inspired by talent and dedication such as we've encountered today and yesterday, it gives us the strength needed to keep fighting against formidable odds. It renews our resolve to continue working towards our goal of ending a tragic situation.

Thank you for all staying with the fight to eliminate cat and dog overpopulation.

# APPENDIX A

# UPDATES

**ANN COTTRELL FREE:** "Hell in Paradise: Vieques, Puerto Rico."

Many of the suggestions made in 1987 to help Vieques' animals have been translated into partial reality: the organization of the Vieques Humane Society; a twice monthly low-cost spay/neuter clinic; an adoption and euthanasia program; a small humane education program; and a 1988 Navy–sponsored two-week inoculation clinic.

Response to the 1987 pleas made this possible. The grant from the Albert Schweitzer Animal Welfare Suggestion Fund (see p. 75) got us off the ground. Individual contributions followed graphic articles in *Animals' Agenda* and the Associated Humane Societies' *Humane News*. The inoculation (against rabies and other diseases) resulted from intervention of the then Defense Secretary Casper Weinberger, whose attention to our pleas was drawn by the Animal Welfare Institute's Christine Stevens. Nearly 1,200 dogs and cats were inoculated. The Humane Society of the United States contributed dog collars. To date, more than 1,000 dogs have been spayed/neutered and a similar number euthanized. Good adoptive homes have been found for some. Humane literature, posters and a pet show have been used to awaken the residents to their responsibilities. Plans are being drawn up to tackle the free-roaming horse problem. Due to Navy and local government indecision, little progress has been made in obtaining a shelter or paid shelter personnel. The Vieques Economic Adjustment Program has been unable to help.

Progress was interrupted in September 1989. Hurricane Hugo delivered a devastating blow to Vieques—fatal to many animals (especially free-roaming horses) and destroying close to eighty percent of the dwellings. Emergency sheltering has been provided the people, but their hungry, abandoned animals were at large, with the few hard-pressed humane society people left with the food and veterinary help problem. Main island supporters shipped food as soon as possible, but in relatively small amounts. The need continues.

As mainland representative, I immediately sought aid from the U.S. Navy and humane and animal relief groups, with little real success, save for the interest taken by the American Humane Association, which sent a substantial contribution towards building a shelter. No adequate telephone communication was possible for several months. I sent a postcard appeal to supporters to help meet what will be continuing needs.

I wish I could list those who are giving a helping hand. The animals cannot be aided nor can the Society function without outside help. Checks can be sent to:

>    Vieques Humane Society and Animal Rescue
>    PO Box 1012
>    Vieques
>    Puerto Rico 00765.

**ROBERT CASE:** "What Part Can Veterinarians Play?"

In 1990, we are still not devoting enough time and attention to educating the pet owner.

In connection with another important approach: There is more interest in, and more attention is being paid to, the concept and use of early spay/neuter.

**DR. LEO LIEBERMAN:** "Neutering Sexually Immature Animals"

Since I made the presentation on early-age neuter surgery (spay/castration) of dogs and cats to the New York State Humane Association in New York City in September 1987, there have been a number of developments:

The commentary in the September 1, 1987 *Journal of the American Veterinary Medical Association* has been reprinted and excerpted throughout the English-speaking countries, and many inquiries have come from other parts of the world.

The most important development is the report to the American College of Veterinary Surgeons, in February 1989, by Salmeria and Bloomberg of the University of Florida, on "Prepuberal Gonectomy in the Dog: Effects on Skeletal Growth and Physical Development." It shows that the changes that occur in the bitch spayed at seven *months* of age are insignificantly different from the changes that occur in the litter-mate bitch pups spayed at seven *weeks* of age. This is the world's first controlled study of the most common veterinary surgical procedure.

The ASPCA of New York City has endorsed and encouraged the concept.

The New York State Humane Association Board of Directors has approved the following resolution:

> RESOLVED, *The New York State Humane Association encourages shelters to investigate the neutering, by licensed veterinarians using safe and humane procedures, of puppies and kittens at the age of two to four months, with a view to make it possible to neuter all dogs and cats before they are adopted.*

The Chicago Animal Control Facility now permits the adoption of neutered animals only—100 percent. Some of these have been done as early as six weeks of age. Two clerks who used to phone recalls and make appointments for surgery, have been put on other duties.

Some shelters and their veterinarians have been unwilling to make the "drastic" change to two months, for surgery. They have reduced the age for surgery to three or four months, and plan to reduce it further as surgical confidence is developed.

To convey the early-age spay/neuter surgery in its proper perspective, the choice of terms is now "neuter-at-adoption."

Veterinary practitioners in general have not responded well to this change in their own patient load. However, many have found and reported the advantages of doing surgery on puppies less than the conventional seven months of age. Some have expressed concern about the adverse economic impact, especially if the shelter in-house surgical facility develops into a "full-service" veterinary operation.

The subject of early spay/neuter has frequently brought the animal welfare community and the veterinary association members into the same room to participate in discussion about neuter-at-adoption. It has had a positive effect on a cooperative relationship between the veterinary community and the animal welfare profession.

On the national level, the AVMA has been looking more closely into animal welfare issues, and considering the possible change of some policy positions. One official wrote that he believes "that it is inexcusable for humane societies to adopt out intact animals." Thirty years ago, as a shelter veterinarian, he did neuter-at-adoption.

American Humane Association has been seeking technical advice on the many questions that reasonable people can speculate on as to the long-term effects of this "new concept." A position paper may be anticipated for the future.

Some humane organizations have not approved the concept. Presumably more data will have to be obtained and analyzed before they feel that they can endorse it.

The College of Veterinary Medicine of the Texas A & M transport their senior surgical students 100 miles to the Austin SPCA shelter for training in early-age neuter surgery. The University of Florida is negotiating a similar arrangement.

In the United Kingdom, this concept has been received with enthusiasm in some quarters.

Analysis of a written survey of veterinarians in the Los Angeles area indicates a wide spectrum of opinions, ranging from enthusiastic endorsement to frank opposition. The largest majority have a "wait and see" attitude, until more information is published.

Funding for studies to provide technical answers and follow-up information appears not to appeal to donors. Further technical work is being delayed for lack of funds.

A standardized national identification system also needs investigation.

Appendix A

So far, breeders have rejected the idea of using early-age neutering. They prefer to sell "pet stock" or puppies "without papers."

In general, the animal welfare community finds neuter-at-adoption a desirable tool in their program.

It is expected that, in time, neuter-at-adoption will be universally accepted and perhaps even mandated by law.

As of June 1990, the following organizations are known to be performing early-age spay/castration:

- Denver Dumb Friends League, CO—4 months, planning to go to 2 months
- Broward County Humane Society, Ft. Lauderdale, FL
- Martin County Animal Rescue League, Stewart, FL
- Chicago Animal Control, Chicago, IL
- Angell Memorial Hospital, Massachusetts SPCA, Boston—8 clinics.
- SPCA Serving Erie County, Tonawanda, NY
- Greenville Humane Society, SC
- Memphis Animal Control, TN
- Central Vermont Humane Society, Montpelier, VT
- Lee County Humane Society, Fort Myers, FL
- Jefferson Parish Animal Shelter, Marrero, LA
- Austin/Travis Humane Society, Austin, TX
- Kings County) Animal Control, Seattle, WA
- Los Brazos Animal Shelter, Bryant, TX
- Vancouver, British Columbia SPCA— 5 shelters
- Bishop SPCA, Bradenton, FL
- Brookhaven Animal Control, L.I., NY
- Southern Oregon Humane Society, Medford, OR
- Baltimore Humane Society, Baltimore, MD

## Updates

- Los Angeles Animal Control, CA
- Miami Humane Society, FL
- Miguel Bracho, DVM, Minnetonka, MN
- Royal SPCA, London, England
- Pinellas Animal Control, Clearwater, FL
- Aluchua County Animal Control, Gainsville, FL
- MEOW, Inc, New Fairfield, CT
- Humane Society of Larimer County, Fort Collins, CO

**ELINOR MOLBEGOTT, ESQ.:** "Legislation: Past & Future"

Westchester County has enacted pets-in-housing legislation similar to that which exists in New York City, prohibiting enforcement of no-pet clauses after the tenant has kept a pet for three months or more with the knowledge of the landlord or landlord's agent, and the landlord has failed to commence legal proceedings to enforce the no-pet clause within that three month period.

Bills have been introduced into the New York State Legislature, to:

**1.** Establish a pet overpopulation control fund similar to that in New Jersey which would subsidize the cost of spay/neuter procedures for those people who are indigent or who have adopted their dogs or cats from a pound or shelter.

**2.** Substantially increase the purebred license fee.

**3.** Prohibit laboratories in New York State from using any animal from a shelter or pound outside of New York State. (These laboratories are already prohibited from procuring New York State shelter or pound animals.)

Other bills which relate to the overpopulation problem have been pending in Albany for several years. These include bills to:

## Appendix A

**1.** Require shelters to spay/neuter animals before releasing them for adoption, or require adopters to pay a deposit for the spay/neuter surgery.

**2.** Allow senior citizens to keep pets in public housing projects.

**3.** Prohibit enforcement of no-pet clauses in leases.

## Appendix B

## Editor's Notes

In these notes, the term "neuter" includes surgical sterilization of both males and females.

1. New York State Assembly Bill A.4968-C, introduced by Assemblyman Maurice Hinchey, had a Senate companion bill, Senate Bill S.3581-C, introduced by Senator Charles Cook (Both legislators were honored by NYSHA for introducing the bills.) S.3581-C was defeated in 1988, because of strong opposition by a small group who objected because, among other allegations, they claimed that it "would prevent the adoption of dogs and cats from shelters." It is hard to understand why those people who do not wish to neuter their pets should be permitted to adopt them from shelters at all. Responsible shelters make it a requirement for adoption, that the pet be neutered. Is the adopter more likely to be able to afford the fee later rather than at the time of adoption?

Programs to subsidize neutering are certainly of importance, and in many areas are needed. But they are in no way a *substitute* for mandatory neutering. New York State Assembly Bill A.7412-C, introduced in 1989 by Assemblyman Maurice Hinchey, provides for such subsidy, the funds to come partly from a differential in license fees for neutered and unneutered dogs.

The ASPCA in New York for many years has neutered animals adopted from its shelters, free. The number of those taking advantage of this has been as low as 20 to 30 percent. The ASPCA has also tried requiring a deposit, refundable when the animal is returned for neutering. For the short period for which data is available, the adoption rate fell slightly, but the percentage of adopted animals neutered rose slightly. The policy was discontinued because it was felt that it put the ASPCA at a disadvantage *vis-à-vis* other shelters in the City, whose adoption requirements did not include a sizable spay/neuter deposit. (Personal communication, John Kullberg). This points up the desirability of standardizing spay/neuter requirements

## Editor's Notes

among shelters, as Assemblyman Hinchey's Bill A.1223-A would do.

The fact is that very few shelters have a satisfactory neuter rate. A small percentage of fertile animals is an important source of overpopulation and strays. Unless the neuter rate is close to 100 percent, the higher the adoption rate, the more the overpopulation is increased. (See Note 4)

It is also argued that commercial breeders should be better regulated. No question. The same conditions required for adoption of shelter animals should apply also to breeder and pet shop animals.

Most programs to encourage neutering are based on subsidies to make the surgery less costly. That's fine. But unfortunately, cost may not at all be the most important factor in resistance to neutering by many who adopt a cat or dog. Some of these factors are discussed in several of the conference talks.

It is interesting to examine the information and conclusions of several earlier surveys and conferences on overpopulation:

In an Argus Archives Report (No. 4, 1973, *Unwanted Pets and the Animal Shelter*), shelters were asked what they considered specific factors affecting the stray problem. One of the fourteen shelters replied: "The fact that we do not have a spaying program and low income people cannot afford to do this." The report states that: "The need for reasonable spay/neuter clinics has been identified almost unanimously by shelter personnel interviewed as the top priority to help reduce overpopulation. . . ." But in answer to the question: "What do you think are the reasons people bring in unwanted pets?" a reply was: "Animal never spayed." The shelter felt that the owners are "usually proud to have pups or kittens."

The Argus Archives Report also quotes a recommendation of the British organization FRAME (Fund for the Replacement of Animals in Medical Experiments, Eastgate House, 34 Stoney Street, Nottingham, NG1 1NB, England): "Only authorized breeders should be licensed to raise and sell domestic pets to the public, and only males or spayed females be permitted for sale."

As reported in the *Proceedings of the National Conference on Dog and Cat Control in 1976* (sponsored by AHA, American Kennel Club, AVMA, HSUS and the Pet Food Institute):

## Appendix B

Close to 80 percent of the people, both dog and cat owners and non-owners, "favor controls requiring that more dogs and cats be neutered in order to prevent unwanted litters."(p. 31)

A report on a reduced fee spaying program of the Southern California Veterinary Medical Association (p.201) states that:

> *During the one year of operation, the communication centers received thousands of calls. However, almost 95% of these calls were requests for services other than surgical sterilization. This overwhelming majority of callers were adamant in their disinterest in having their pets spayed or neutered —even though the service was available at no cost. In many cases the pet owner became resentful and expressed displeasure at the suggestion that the animal undergo surgery to prevent pregnancy.*

Clearly, while most people favor neutering, there are those who do not, and will not be influenced by its availability at a reduced rate, or even free. If 20 percent of the population that adopts cats or dogs do not favor neutering, and shelters allow unneutered animals to be adopted by these people, then the persistence of overpopulation is assured. It is also true that for many people, an investment in the animal will motivate more responsible ownership. *Only mandatory neutering will be effective.* And if neutering is made mandatory, provision must be made to make it affordable to all. Seven states now have mandatory spay/neuter laws: Arkansas, California, Florida, Illinois, Kansas. Massachusetts and Oklahoma. It is predictable that the efficacy of these laws will depend on proper enforcement, including an adequate deposit if animals are released before they are neutered.

In the *Proceedings of the First New England Conference on Animals and Society*, Tufts University School of Veterinary Medicine, 1984, Andrew Rowan points out that: ". . . there is no consistent pattern to shelter population trends after starting a spay/neuter program . . . we still lack sound information on the reasons why people do or do not have their pets altered and the importance of financial considerations."

In the *Proceedings of the Conference on Animal Management and Population Control*, Tufts University, 1985, Diane Quisenberry of the Animal Control Division, Charlotte, NC, reports that the current animal control program, instituted in 1982, consists of five parts:

## Editor's Notes

1. *A low-cost Spay/Neuter Clinic, which is available to all citizens.*

2. *Mandatory spay/neuter for all animals adopted from the shelter.*

3. *Differential licensing fee for dogs and cats.*

4. *A strong ordinance with stringent enforcement.*

5. *An on-going education and advertising program.*

In 1978 the Animal Control Division instituted mandatory spay/neuter of all animals before adoption, and a low-cost spay/neuter clinic. There was an initial increase in the number of euthanasias, which slowly levelled off over the years. The decrease in the adoption rate is viewed as a *good* development, since it screens out uncommitted owners.

Quisenberry reported:

> *The trend of general public use of the Spay/Neuter clinic has decreased over the two year operation. In the same way, the number of mandatory spay/neuters from the shelter (i.e. number of adoptions) has increased over the period.*

When the mandatory spay/neuter legislation was first passed, adoptions fell. (Adoption fees included the price of the spay/neuter.) "This change was expected," said Quisenberry, "and even encouraged, since the higher prices became an excellent screening tool.... Over the last two years, the adoption rate has slowly increased." Since 1985, the adoption rate has continued to increase, even when the adoption fee was made somewhat higher. (Personal communication from Diane Quisenberry's office.)

Quisenberry further reported:

> *The mandatory spay/neuter program in Charlotte may be a little different from similar programs in that we require that adopters of adult dogs and cats purchase their animal and pick it up two days later, after the spay/neuter surgery has been performed. There have been surprisingly few complaints about this requirement.... By having the program set up in this way, 100% of the adult animals are surgically sterilized before leaving our custody.*

## Appendix B

From the NYSHA conference, we have some clear ideas of why some people do not want to have their pets neutered. *These reasons will not be affected by financial considerations. They can be dealt with only by education and by mandatory neutering.*

**1a.** See *Legislative Resolution* before opening of conference. Senator Charles Cook and Assemblyman Maurice Hinchey received tributes from NYSHA for introducing this resolution to the New York State Legislature.

**2.** Good animal control involves both adequate training for animal control officers, and good shelters. The first problem is beginning to be addressed by providing schooling for animal control officers. But we do not yet have proper requirements for *qualification* as an animal control officer. Laws are urgently needed also, which would set shelter standards. This is more and more being demonstrated by conditions found on the premises of a number of animal collectors, including the Bridgehampton shelter, closed in 1985, and the Animals Farm Home in Ellenville, closed in 1988. It is expected that a shelter standards bill will be introduced into the New York State Legislature in 1991.

**3.** The problem of adoption of animals by transients, including students and summer residents, is a widespread one. A little "humane education" would seem to be in order at colleges and universities—perhaps literature given to every student. Also, it would be necessary to know what provision is being made for the animal between semesters, and after the student graduates. As for summer resorts, it is questionable whether animals should be adopted out, except to permanent residents.

**4.** Pet overpopulation is due to failure to neuter. It seems clear that the *responsibility* lies with all those who adopt out, sell or keep as pets, animals that are not neutered.

Only licensed breeders should keep breeding stock. All animals they sell should be neutered.

Shelters should adopt out only animals that are neutered.

## Editor's Notes

Pet shops should sell only neutered animals.

At present, neutering of puppies and kittens is not generally practiced. But see presentation by Leo Lieberman on early neutering.

Shelters strive for a high adoption rate, in part because it sounds good in fund-raising, and partly because they sincerely wish to give as many animals as possible, a good life. Unfortunately, almost every adopted animal that is not neutered, increases the number of strays. NYSHA has suggested that shelters report, instead of "adoptions," "effective adoptions," defined as the number of adoptions, minus the number of adopted animals not neutered, minus the number of their progeny within the next two years. *Unless the effective adoption rate is a positive figure, the lower the effective adoption rate, the greater is the potential for population increase or for increased numbers of deaths.* If we calculate the *effective adoption rates* for *adoption rates* with the same percentage of adopted animals neutered, it becomes apparent that the higher the adoption rate, the lower will be the effective adoption rate. To put it differently, unless all adopted animals are neutered, the higher the adoption rate, the more unwanted animals are produced. This is easily calculated:

If a shelter adopts out 20 out of 100 animals, the "adoption rate" is 20 percent. If 90 percent (18) of these are neutered, there will be two unneutered animals released. These may have two litters of, say, 8 pups each during the year (32 pups). It is highly unlikely that those who do not neuter their own animals will ensure the neutering of the progeny of these animals. If 16 of these 32 two pups each has a litter of 8 within the next year (128 pups), the "effective adoption rate" is the original 20 dogs adopted out, minus the original 2 unneutered animals (equals plus 18), minus the 32 pups (equals minus 14), minus their progeny of 128 pups (equals minus 142). If we add to this the number of animals originally euthanized (80), this means 222 animals are "in excess."

If 40 of the original 100 shelter animals are adopted out, the "adoption rate" is 40 percent. If 90 percent of these animals are neutered, there will be 4 unneutered animals released, each may have two litters of 8 pups each during the year (64 pups). If 32 of these have litters of eight, 256 more pups will be produced. The "effective adoption rate" would be the number of adoptions (plus 40) minus 4 unneutered animals

## Appendix B

(equals plus 36), minus their progeny, 64 pups (equals minus 28), minus *their* progeny of 256 pups (equals minus 284). If we add to this the number of animals originally euthanized (60), the total of animals "in excess" is 344.

These figures do not even take into account the numbers of animals born in future years. It can easily be seen that even when 90 percent of the adopted animals are neutered, the shelter may well be responsible for adding to, rather than reducing the stray population. *In fact, unless the "effective adoption rate" is a positive figure, then the higher the adoption rate, the higher the resulting number of unwanted animals.*

While some unneutered adopted animals may not reproduce, or may be returned to the shelter, or may die or be euthanized before reproducing, it seems clear that it is the percentage of *neutered* animals adopted rather than the *total number* of animals adopted, that will, in the long run, influence the number of strays and the number of animals that will die.

It has been argued that only a small percentage of dogs and cats come from shelters. There is no way of knowing: the parents or grandparents of puppies and kittens supplied by the neighbor or picked up on the street, may have been adopted as shelter animals that were not neutered. Legislation is needed so that, eventually, no one except a licensed breeder could keep an unneutered animal.


5. The rate of "Returned-to-Owner" undoubtedly varies greatly from shelter to shelter. It is generally lower than it should be, and would undoubtedly be increased by using a computer. However, there are many reasons why more animals are not returned to their owners. One important reason is that the owner does not want the animal and has either deliberately abandoned it, or will not bother to claim it. Tattoo of all animals adopted from shelters would serve both to help return lost animals, and to locate irresponsible owners who have abandoned animals. It might even discourage abandonment (which, in New York State, is illegal). In addition, a tattoo done at the time of surgery, indicating that the animal was spayed, would avoid uncertainty. This is suggested by speakers in the conference, and by members of the audience.

Insertion of a microchip is another method that is now possible.

Editor's Notes

**6.** The ASPCA still has a full-service shelter in Manhattan, and in Brooklyn. An active group, Bronx Animal Rights Coalition (BARC), has been "getting people to ask" the City to provide at least a "receiving facility" in the Bronx. In their Oct/Nov 1989 newsletter, they reported that it was expected that a Pet Receiving Center for the Bronx would probably be opened at the beginning of 1990. This was delayed.

Staten Island still does not have a shelter, or even a receiving facility.

According to a Sept. 24, 1989 article in the Real Estate Section of the *The New York Times,* the ASPCA: "is to float $22.8 million in revenue bonds with the help of New York City to help pay for the largest expansion of its facilities in the metropolitan area in its 123-year history." The new headquarters will be at 424 East 92nd street (about one block from the present headquarters). The proceeds from the bonds will also be used to build a new two-story shelter at 330 East 110th Street.

In addition, "the society is considering building a new veterinary hospital in Brooklyn, and a new shelter in Queens at Kennedy International Airport...."

Dr. Kullberg, announced in June 1990 that the ASPCA has reached an agreement with the City, which calls for a total operating budget of $6,500,000, of which the City will fund $5,050,000 of the operating budget for the coming fiscal year. As part of the agreement, the ASPCA is committed to construct and pay for a new $6,000,000 shelter in Manhattan. The ASPCA also plans to continue operation of its shelter in Brooklyn, and its three "Pet Receiving Centers" in Queens, Bronx and Staten Island.

**6a.** The National Association for the Advancement of Humane Education underwent a name change in 1989, and is now known as the National Organization for the Advancement of Humane and Environmental Education (NAHEE). Excellent humane education materials can be obtained from NAHEE, whose mailing address is: P O Box 362, East Haddam, CT 06423

**6b.** The video program, *The Control of Feral Cats,* is available for $33.75, from:

Appendix B

Universities Federation for Animal Welfare (UFAW)
8 Hamilton Close
South Mimms
Potters Bar
Herts EN6 3QD, England

**7.** Tomahawk traps are available from:

Tomahawk Live Trap Company
P.O. Box 323
Tomahawk, WI 54487
(715) 453-3550

**8.** Assembly Bill A. 5067-A (Senate Bill S.3410-A) had two unrelated provisions: (1) It ended the use of animals from pounds in New York State, for research and (2) it banned certain methods of killing animals in shelters and by dog control officers, and made sodium pentobarbital available to shelters.

New York State Senators Joseph Bruno and Frank Padavan, and Assemblyman Arthur Kremer received NYSHA legislative awards for their sponsorship of this measure.

In regard to the use of pound animals, the problem remains that laboratories may still use animals obtained from pounds in other states. Legislation has been introduced to correct this situation. Bill S.1669, sponsored by Senator Joseph Bruno and Bill A.2584, sponsored by Assemblyman Ivan Lafayette, would ban the sale in New York State of animals from pounds in other states. NYSHA strongly supports such legislation.

In regard to euthanasia, the problem remains that the best injectable euthanasia agents are humane only if they are injected by a person skilled in the procedure. The Bureau of Controlled Substances of the New York State Department of Health, has promulgated regulations under this new law, stating who may administer intravenous barbiturate injections, and the responsibilities of the person accountable for the barbiturates in the shelter. Under the auspices of the New York State Animal Control Association (NYSACA), and the New York State Humane Association (NYSHA), meetings have been held to explain the regulations, and courses are being offered to shelter personnel wishing to qualify for administering intravenous injection of barbiturates.

## Editor's Notes

Although the law provides that shelters can obtain sodium pentobarbital, the regulations that have since been promulgated under the law provide that sodium pentobarbital may be obtained by shelters only as a solution. "Solution" is defined as: "a premixed solution of sodium pentobarbital, manufactured specifically for the euthanasia of animals, which contains such other ingredients as to place such solution within Schedule III of the Controlled Substances Act. (Article 13, PHL)"

**9.** An excellent film, *Helping Animals in Israel,* is now available from CHAI, PO Box 3341, Alexandria, VA 22301.

**9a.** An eloquent expression of this attitude appeared in a letter printed in Phyllis Wright's column in *Shelter Sense* (HSUS publication). Speaking of humane euthanasia technicians, it says:

> *They do not dwell on the fact that a thin, hairless stray will undoubtedly be passed up for adoption and be put down after five days. They realize that this situation is beyond their control, for each day they are faced with an astronomical number of thoughtlessly produced and unwanted animals. They set immediately to the task of freeing this poor animal, possibly for the first time, from the torment of fleas and dirt via a warm, relaxing bath. They follow this with a walk in the sun and days of cookies, sympathetic words, and friendly scratches in just the right spots. And when that fifth day arrives, this old dog's final thoughts are of being warm and secure in the arms of a trusted friend, of feeling sleepy, and then of nothing.*
>
> *What makes these exceptional people stay in what would be to many of us an emotionally devastating situation? I maintain that it is a great love, greater than most of us are able to summon... 'You work at the humane society? Oh, I could never do that. I love animals too much.' I have come to understand that 'too much' is in fact not enough.*

While this description of shelter euthanasia is, for many facilities, certainly idealized, the concept of caring, humane euthanasia is valid, as is the existence of deeply concerned people who carry it out.

## Appendix B

**9b.** See the presentation by Wolfgang Jöchle.

**10.** Lifeline for Wildlife closed in 1989. It is unknown whether it will reopen. There is no comparable rehabilitation center in the area, and one is desperately needed. It would seem appropriate for state and local governments to support such centers.

There is also a need for more individual rehabilitators, as well as for people who would provide transportation, so that injured wildlife might conveniently be brought to a licensed rehabilitator.

**11.** It has been suggested that a license should be required for owning a dog or cat (*Proceedings of the National Conference on Dog and Cat Control*, 1976, p. 262)

> *Organizations sponsoring the National Conference on Dog and Cat Control, along with city and county governments, humane organizations and animal control agencies, should study and consider the potential merits of an animal control system requiring an individual to obtain a permit or license to own a pet rather than a license for the pet itself.*

**11a.** There should also be a legal requirement that such animals be neutered when they are sold.

**12.** More and more, responsible scientists are recognizing the social as well as the scientific problems with the use of pound animals.

At the 1979 session of the New York State Senate during which the Metcalf-Hatch law was repealed, Senator Padavan, who led the fight for repeal, read the following statement signed by about 25 scientists, including several prominent names:

> *Metcalf-Hatch is an ill-conceived law, damaging to the good name of science and to its quality. The use of animals from shelters for experimentation is not only unnecessary and unethical, but is detrimental to sound research.*
>
> *Strays are of undetermined genetic and environ-*

## Editor's Notes

> *mental background and react unpredictably and inconsistently, making questionable the reliability of most research in which they are used. Metcalf-Hatch perpetuates inferior research.*
>
> *The use of shelter animals for experimentation creates a schism between pet-owners and research scientists. Further, it equates the interests of scientific research with the permanent existence of a large population of strays—a concept repugnant to the scientist as a decent human being.*
>
> *Metcalf-Hatch is a bad law. Its repeal should not be delayed.*

During the years since then, The Fund for Animals has collected the signatures of several hundred scientists opposing pound seizure in parts of California. And five separate groups of scientists concerned with experimental animals have been formed: Association of Veterinarians for Animal Rights (AVAR), Medical Research Modernization Committee (MRMC), Physicians Committee for Responsible Medicine (PCRM), Psychologists for the Ethical Treatment of Animals (PsyETA), and Scientists' Group for Reform of Animal Experimentation (SGRAE). All of these groups are opposed to pound seizure.

**13.** The idea that animals are here on earth purely for the benefit of man is a very old one.

James Serpell (*In The Company of Animals*, p. 123) explains that according to Aristotle, animals were here only to serve the purposes of man. Cruelty to animals was to be avoided, but only because it might encourage cruelty to human beings.

Aristotle also argued that the sun and the planets rotated around the earth.

Both ideas were incorporated into early Christian doctrine.

**14.** As a result of staff cutbacks, Bide-A-Wee's humane education program was virtually discontinued in 1989.

**15.** This is a valid argument for the existence of "no-kill shelters." It is important that shelters that do euthanize recognize that there is a place for a no-kill shelter, provided that it is not the only one in the area. Because a no-kill shelter cannot take in all animals brought to it, it is important there be a shelter in the area that does euthanize, so that it can accept all animals brought to it.

**16.** The co-operation described here is a valuable one.

The "battle" between shelters that euthanize, and "no-kill" shelters is based partly on the assertion by some no-kill shelters that: "We do not believe in killing animals," and on the advantage that this gives them in public relations and in fund-raising. Shelters that *do* euthanize, feel that they are left with all the "dirty work." Both types of shelter should recognize the necessity for, and the value of, the other type:

No-kill shelters can say that they do not kill animals, but should *never* say that they *do not believe* in killing animals.

Shelters that *do* euthanize should accept that there *is* a place for shelters that do *not* euthanize, recognizing that certain individuals would under no circumstances take an animal to a shelter where euthanasia might be performed.

Both types of shelter have the responsibility of educating the public on the problems of overpopulation.

**17.** The availability of a fenced-in yard, should be viewed with extreme caution. While it *can* be a great advantage, it is unsafe unless the dog is never left out unsupervised. Too many dogs have been lost from "secure" fenced yards. In my opinion, the assurance that "We have a fenced-in backyard," should be regarded as a danger signal, and carefully investigated.

**18.** Massachusetts law not only prohibits the use of animals from pounds in the state, but prohibits the importation into the state, of pound animals from any other state, for purposes of research.

Other states that ban pound seizure include Connecticut, Delaware, Hawaii, Maine, Maryland, New Hampshire, New Jersey, New York, Pennsylvania, Rhode Island, South Carolina, Vermont and West Virginia.

Editor's Notes

**19.** When the Metcalf-Hatch Act was passed, in 1952, the ASPCA went along with the new law, against the clearly expressed wishes of its members. Twenty years later, the ASPCA sent a communication to its members, which stated, in part:

> Since its inception and passage in 1952 the law requiring public pounds to relinquish animals to laboratories for scientific purposes has caused the American Society for the Prevention of Cruelty to Animals much concern, to be regarded in the minds of many of the public as less than humane and in general has been somewhat of a deterrent to the good operation of the Society. . . . Regardless of anything to the contrary, the facts and philosophies above have to a great degree, hampered the operations of the Society these past 20 years. . . . Inasmuch as section 505 of the Metcalf-Hatch Act not only fails to provide support but acts to our disadvantage then the time has come to change the law through the means of repeal of that section.
>
> The support of our members and benefactors is needed to make that change.

In 1979, the coalition to Abolish Metcalf-Hatch, spearheaded by Henry Spira, was formed. About thirty humane organizations (including the New York State Humane Association) were listed on its letterhead.

**20.** Stolen animals are frequently sold to dealers at auctions. A 1988 U. S. Senate Bill, S.2353, would have prohibited dealers from selling animals acquired at auctions, to laboratories. But the language of this bill was such that it seemed to legitimize pound seizure by listing shelters and breeders as approved sources. Part of the bill read:

> Except in instances where State or local law supersedes it shall be unlawful for any class B licensee . . . to obtain live random source dogs from a source other than:
>
> (1) a state, county or city owned pound or shelter
>
> (2) a private entity established for the purpose of caring for animals such as a humane society. . . .

## Appendix B

*(3) research facilities licensed by the Department of Agriculture*

*(4) individuals who have bred and raised such dogs and cats on their own premises.*

Scientific groups opposed the bill because they did not want to limit their sources of experimental animals.

Humane groups were split on support of the bill. It did prohibit the purchase of dogs and cats at auctions, which not only deal in stolen animals, but subject the animals to be sold to inhumane conditions and treatment. Some humane groups supported the bill for this reason.

On the other hand, certain humane groups opposed it because the language as to the use of pound animals was dangerously unclear. They feared that it might set back the gains made against pound seizure, and interfere with obtaining anti-pound seizure legislation in more states.

Without pound seizure, purpose-bred animals would be the only source of experimental animals. This would undoubtedly cut down on the numbers used.

**21.** If the time came when there were no unneutered pets, a different source of support for subsidized spay/neuter would have to be provided. It is a proper function of the State to lend support for spay/neuter, as a function of animal control. This expense should not fall on the shelters. But it is highly doubtful that either differential licensing or a nominal spay/neuter fee, or both together, will result in approaching 100 percent spay/neuter. This will happen only if spay/neuter is mandatory. In New Jersey, the percentage of licensed dogs neutered rose from 47 percent to 57 percent from 1985 to 1989. The percentage of dogs licensed is not known.

**22.** The pertinent section (809) of the New York State Education Law reads:

*The officer, board or commission authorized or required to prescribe courses of instruction shall cause instruction to be given in every elementary school under state control or supported wholly or partly by public money of the state, in the humane*

## Editor's Notes

> treatment and protection of animals and birds and the importance of the part they play in the economy of nature as well as the necessity of controlling the proliferation of animals which are subsequently abandoned and caused to suffer extreme cruelty. Such instruction shall be for such period of time during each school year as the board of regents may prescribe and may be joined with work in literature, reading, language, nature study or ethology. Such weekly instruction may be divided into two or more periods. A school district shall not be entitled to participate in the public school money on account of any school or the attendance at any school subject to the provisions of this section, if the instruction required hereby is not given therein.

**23.** This should certainly be true. Obviously, if all animals adopted from shelters were neutered, more neuterings would be done. But presumably, cooperation of veterinarians would be needed, as some neuterings, at least, would have to be subsidized.

**24.** Humane education, especially for children, is effective, in my opinion, only when it arouses empathy. The films listed all satisfy this requirement. The one shown illustrates the point.

**25.** Kellert, S.R. and Felthaus, A.R., "Childhood Cruelty Toward Animals Among Criminals and Non-criminals," *Human Relations*, Volume 38, Number 12, pp. 1113-1129, 1983.

**26.** This humane education program is distinctive in being one of the very few that concentrates on adults rather than children.

**27.** See Note 13. It is encouraging that religious doctrine can evolve and progress. A recent report of a committee of the World Council of Churches (reported in *The Animals Agenda*, April, 1989) discusses the manifold cruelties to which animals are subjected, and recommends in part:

> *In view of the widespread maltreatment of animals*

## Appendix B

*throughout the world and in view of the intrinsic value of individual animals to themselves and to God, we recommend that the subunit on Church and Society of the World Council of Churches take appropriate steps to:*

*a. Encourage the churches and their members to acquire knowledge about how animals are being treated and in what ways this treatment departs from respect for the intrinsic value to themselves and of animals as creatures of God.*

*b. Encourage the Christian community to consider actions such as:*

> *1. Avoid cosmetics and household products that have been cruelly tested on animals. Instead, buy cruelty-free items.*
>
> *2. Avoid clothing and other aspects of fashion that have a history of cruelty to animals, products of the fur industry in particular. Instead, purchase clothes that are cruelty-free.*
>
> *3. Avoid meat and animal products that have been produced on factory farms. Instead, purchase meat and animals products from sources where the animals have been treated with respect, or abstain from these products altogether.*
>
> *4. Avoid patronizing forms of entertainment that treat animals as mere means to human ends. Instead, seek benign forms of entertainment, ones that nurture a sense of the wonder of God's creation and reawaken that duty of conviviality we can discharge by living respectfully in community with all life, the animals included.*

We can only hope that eventually the churches will address also the problem of overpopulation of companion animals—a problem not dealt with in this report.

**28.** Apropos this question, an interesting editorial in the November 1989 *Animals' Agenda* includes the following paragraphs:

## Editor's Notes

*"Humane" is a word that has fallen from grace these days among animal defenders who prefer their work described as "animal rights" or perhaps more clearly, "animal liberation." The word "humane" is considered old-fashioned and is associated with the "animal welfare" school of philosophy which directed people not to stop using or killing animals but to do it kindly. But while it has direct implications for animals, "humane" is an adjective used to describe civilizing human activities or the better qualities of human nature. It is a word not to be despised.*

*Whether we call our work and our movement "animal rights," "animal liberation" or "humane," our efforts have extremely broad implications for human morality. The movement that attempts to raise the status of animals and improve their treatment is, in reality, a human potential movement. For if we succeed in transforming human behavior toward animals, humankind can pass its "test" and move on to a new level of consciousness.*

The valuable suggestions, skills, and untiring work of Samantha Mullen in arranging the format for this volume are gratefully acknowledged by the editor and by the Board of Directors of the New York State Humane Association.

# NEW YORK STATE HUM